Driving Project, Program, and Portfolio Success

The Sustainability Wheel

Driving Project, Program, and Portfolio Success

The Sustainability Wheel

Richard Maltzman
David Shirley

Winners of the 2011 David I. Cleland
Project Management Literature Award

CRC Press
Taylor & Francis Group
Boca Raton   London   New York

CRC Press is an imprint of the
Taylor & Francis Group, an **informa** business
AN AUERBACH BOOK

CRC Press
Taylor & Francis Group
6000 Broken Sound Parkway NW, Suite 300
Boca Raton, FL 33487-2742

© 2016 by Taylor & Francis Group, LLC
CRC Press is an imprint of Taylor & Francis Group, an Informa business

No claim to original U.S. Government works

Printed on acid-free paper
Version Date: 20150729

International Standard Book Number-13: 978-1-4987-0328-4 (Hardback)

Visit the Taylor & Francis Web site at
http://www.taylorandfrancis.com

and the CRC Press Web site at
http://www.crcpress.com

Writing a book could be considered a project, except that without proper discipline it may not have that *definitive end* that is supposed to be characteristic of a project. So, I'd like to start by thanking my coauthor David Shirley for ensuring the book's quality content and schedule. As usual, it was a pleasure to collaborate with Dave. Without Dave, the book may not have ended, but without the support and encouragement of my dear wife Ellen, my daughter Sarah, and son Daniel, the book would never have been started, so it's with much love that I dedicate this book to them. And, although she sometimes (literally) consumed a few pieces of draft material for the book as she went through puppyhood, I also want to acknowledge the companionship and "opinions" of our Beagle/Brittany-mix, Maisie, who often reminded me that writing a book requires hound-like determination.

Richard Maltzman

First, I'd like to thank my coauthor Rich Maltzman, without whose creativity I doubt if this book would ever have been written. He could see the vision from a back-of-an-envelope sketch to a complete book. My creative process involves walking with a companion. In this case my walking companion is my Golden Retriever Odin. I'd like to thank him for listening to my ideas on our walks across the beaches of Cape Neddick. Most importantly, I thank my wife Judi for all the unconditional support she has given me over the last 36 years and for her incredible work ethic. Her unswerving dedication and perseverance for doing the right thing has set the standard for me.

David Shirley

Contents

FOREWORD   xiii

INTRODUCTION: THE SUSTAINABILITY WHEEL™   xv

AUTHORS   xxvii

CHAPTER 1   SUSTAINABILITY AND SUCCESS   1

The Gear Model and Organizational PM   2

Projects, Programs, and Portfolios   3

The Need to Integrate—Not Add   6

Integrating Sustainability: A Key Trend   8

    #5: Organizations Must Build Bottom-Up Processes to
    Link Project Outcomes to Organizational Strategy   8

    #9: Project Management and Business Strategy Better
    Align to the Benefit of the Organization   8

An Example of Integration: The Sustainability
Breakdown Structure   9

Creating Shared Value   10

Benefits Realization and the Real Definition of "Project Success"   12

Project Success and Improved PM Maturity   15

Strategy, Projects, Programs, Portfolios, and Success   20

You've Had the Power All Along—and Our
"3-Click Challenge"   24

Making the Change to Sustainability Thinking in Projects,
Programs, and Portfolios   26

Change Intelligence at Various Hierarchical
Levels in the Enterprise   33

More about Projects, Programs, Portfolios,
Leadership, and Change   37

CHAPTER 2   THE HUB: THE RESPECT DIMENSION   41
Introduction   41
The Mission Statement   42
  Ten Things to Look For in a Good Sustainability
  Mission/Vision Statement. Is it: 44
    Patagonia—Case Study   44
    Stonyfield Farms—Case Study   46
    General Motors—Case Study   47
  EarthPM   49
    EarthPM's Mission and Objectives   50

CHAPTER 3   THE SPOKES: THE REFLECT DIMENSION   53
Introduction   53
Environmental Management Plan   54
EMS and ISO 14001   55
  Enterprise Level   57
  Portfolio Level   59
  Program Level   60
  Project Level   61
Sustainability Programs and Incentives   63
  "Ray"sing the Bar for Sustainability   68
    Why the Program Was Started: An Epiphany
    at the Top of the Company   68
    How the Program Works: It's Part of Every
    Employee's Job   68
    Key Lessons   69
    Tangible Results: Interface Employees
    Neutralize Personal Travel Emissions   70
    Measuring Results: Material Usage and Employee
    Engagement   71
    Challenges: Addressing the "Big" Questions   71
    Advice for Others: Culture Change is Key   72
    Links to More Information   72
  Making Sustainability Personal   72
    Why the Program Was Started: By Wal-Mart
    Associates, for Wal-Mart Associates   73
    Key Lessons   73
    PSP: Program Design   74
    PSP Implementation   75
    PSP Results: A Truly Grassroots
    Sustainability Movement   76
    Tangible Results of PSP Program: Employees
    Making a Difference   76
    Challenges: Keeping it Fresh   77
    Next Steps: International Expansion   77
    Links to More Information   79

MAPing a Route toward Sustainability   79
  Why the Program Was Started: Engaging
  Employees in the Sustainability Mission   80
  How the Program Works: Long-Term Goals Linked
  to Job Performance Measures   80
  Measuring Results: Gains in Key Areas   81
  Challenges: Staying Focused   81
  Next Steps: Seeking Step Change   82
  Key Lessons   82

CHAPTER 4 THE TIRE   85
Introduction   85
Connect   85
  DJSI   86
  Claremont-McKenna's Roberts Environmental
  Center Pacific Sustainability Index (2006 through 2013)   88
  GISR (Global Initiative for Sustainability Ratings)   89
  CSRHub.com   91
  Sustainability Leadership Report: Brandlogic and CRD
  Analytics   91
  Newsweek Green Rankings   93
  ClimateCounts.org   96
  Global 100   98
Detect   101
  Predators   103
  Parasites   103
  Idle Stakeholders   103
  Shy Fans   104
  Confident Fans   104
  Followers   104
  Supporters   105
  Advocates   105
  Champions   105
  Conclusion   106
Reject   113
  Introduction   113
Case Study: Subaru of Indiana Automotive   114
Case Study: AT&T   123
Project   124
  Introduction   124
  Maturity Models   124
Opportunities and Challenges   126
  Business Cases   126
  Environmental and Sustainability Education   126
  Profit   128
  Partnering   128
  Strategic Value Creation   129

Case Study: Walkers Crisps 132
Case Study: Shell Oil 132
Case Study: Marks & Spencer–Beyond Plan A 133

CHAPTER 5   THE ROAD   135
Introduction 135
Dialect 135
   Communications 136
   Tools 137
Intellect 141
   Are We Doing Research on Companies Similar
   to Ours for Their Sustainability? 142
   Are We Using Benchmarking for the Right Reasons? 143
   Are We Sufficiently Educated in the
   Benchmarking Process? 144
Circumspect 144

CHAPTER 6   INTERPRETING THE SUSTAINABILITY WHEEL   147
Initial Feedback 147
Interpreting the Sustainability Radar™ Signatures 147
The Signatures 148
   Leader (Strong in All Dimensions) 148
     What It Means 149
     What to Do 149
   Laggard (Weak in All Dimensions) 149
     What It Means 149
     What to Do 150
   Theorist (Weak in Connect, Reflect, and Reject) 150
     What It Means 150
     What to Do 151
   Greenwasher (Weak in All Dimensions except Connect) 151
     What It Means 151
     What to Do 152
   Exploiter (Weak in All Dimensions except Project) 152
     What It Means 153
     What to Do 153
   Drone (Weak in Respect) 153
     What It Means 154
     What to Do 154
   Efficient Bamboozler (Weak in Respect, Reflect) 154
     What It Means 154
     What to Do 155
   Inefficient Optimist (Weak in Reject, Detect) 155
     What It Means 155
     What to Do 156
   Shy Pessimist (Weak in Project, Connect) 156
     What It Means 157
     What to Do 157

Shy Optimist (Weak in Detect, Connect) 157
   What it Means 158
   What to Do 158
Unmoored Efficiency Expert
(Weak in Detect, Respect, Project) 159
   What It Means 159
   What to Do 159
Efficient Automaton (Weak in Connect, Detect, Respect) 160
   What It Means 160
   What to Do 160
Pilotless Altruist (Weak in Respect, Project, Reject) 161
   What It Means 161
   What to Do 161
Carefully Inefficient Pilot
(Weak in Project, Reject, Reflect) 162
   What It Means 162
   What to Do 162
Inefficient Pessimist (Weak in Project and Reject) 163
   What It Means 163
   What to Do 163
Planeless Pilot (Weak in Reflect) 164
   What It Means 164
   What to Do 165
Shy Drone (Weak in Respect, Reject, Connect) 165
   What It Means 165
   What to Do 166
Operator (Weak in Project) 166
   What It Means 166
   What to Do 167
Theoretical PM (Weak in Connect and Reflect) 167
   What It Means 167
   What to Do 168
Fearless Leader (Weak in Detect) 168
   What It Means 169
   What to Do 169
Pessimistic Planeless Pilot
(Weak in Reflect, Reject, Project) 170
   What It Means 170
   What to Do 170
Sustainability Wheel Pilot Results 171
   Global, Well-known IT Leader 171
   Consultancy Services Enterprise 172
   Design, Construction, Engineering Firm 173
But Wait, There's More… 174

INDEX 177

Foreword

The dissemination of innovation is always a challenge. We know far more than we apply, summarized as the knowing–doing gap. As a practicing social scientist specializing in leading change and transformation, I have had many opportunities to see the best ideas go unused because of lack of attention to the human factor. This human factor is often left out because it is hard to measure and is messy to navigate through. People are amazingly innovative in avoiding doing new things, partly because change "hurts" their brains, requiring the establishment of new neural pathways and using precious attention to adapt to new patterns.

For the last six years, I have focused my change efforts on bringing sustainability focus and engage the behavior change it requires to large groups of employees at Wal-Mart, Frito, AT&T, and Austin Bird. It has been exciting to see thousands of people take up personal practices, which support social, economic, cultural, and environmental sustainability efforts. These efforts incorporated learnings from positive psychology and communication to go beyond shame and blame, focusing people on what makes them happy and then connecting their efforts to actions that have a personal meaning to their lives. Addressing the personal threats to personal autonomy, sense of certainty, and disruption of relationships—before heading into planning and execution—will save time and leave people more renewed at the end of a project, and truly ready for another.

For the past five years, I have been a core faculty in sustainable leadership at Presidio Graduate School, where I have the privilege of working with MBA and MPA students who are committed change agents. The tools and models offered in this book draw on the theories/approaches we use to develop the personal and organizational change capacity.

I have watched Rich and Dave develop this book with the focus on project managers, with excitement, because they are the pivot point in translating new ideas into action and often the first to experience the normal resistance that our brains are hardwired to exhibit when a novel way of doing something is approached. Equipping project managers with the increased insight and well-developed tools/models to approach the reduction of human process loss that often affects projects will reduce the cycle time of implementation and release full commitment to action.

They have built a strong bridge between project management and the introduction of sustainability focus, giving access to critical teams, where thinking about interactive relationships can yield greater impact and momentum for sustainable change. To create an environment where nine billion people can thrive requires project managers who take a sophisticated approach to engaging themselves and the people they work with to do new things, in new ways. I welcome project managers into the league of change agents, who will affect the ability of organizations to be better stewards of our environmental and human resources.

Cynthia Scott, PhD MPH
Core Faculty
Presidio Graduate School
San Francisco, California

Introduction: The Sustainability Wheel™

Rubber Meets the Road

When we wrote *Green Project Management* in 2010, we aimed it at project managers. We discovered, through our research and inspiration from books like *Green to Gold* (Esty and Winston 2009) and *The Necessary Revolution* (Senge et al. 2010) and heroes like Ray Anderson of Interface/FLOR and Steve Howard of IKEA, that business appears to have not only bought into integrating "triple-bottom-line" (we'll call it *sustainability*) thinking into their business plan but also have started profiting from it–see MIT and Boston Consulting Group's *Sustainability's Next Frontier* (Kiron et al. 2013). These books and studies are outstanding. However, they had little or no mention of projects, project management, or project managers. Similarly, it confounded us that *project managers*, those who bring the ideas of enterprise to reality, are seemingly unaware of the level of buy-in being given to sustainability by their own enterprise's leaders, and as a result, their project charters (a source of power and authority for the PM) and plans (the engines that runs their projects) often have little or no mention of sustainability.

In other words, there's a huge disconnect between what we call "the rubber" (ideas, mission, vision, values, and strategy) and "the road" (the day-to-day operations of an enterprise).

For what we can only call *geometric* reasons, let's flashback to 1999 for a moment. In that borderline millennial year, the authors had just completed a presentation to The Conference Board in Mexico City on the relationship between project and quality management. After the presentation, we decided to head to Teotihuacán to climb the 248 steps and 170 steps of the Pyramids of the Sun and Moon, respectively, and to sample tequila. Luckily, we did it in *that* order (climb first, drink afterward). So, we can say, with the benefit of physical exertion if nothing else, *we know our pyramids.*

Now let's flash forward to the present day. Here we were, stymied by the lack of buy-in by project managers to integrating sustainability thinking into their projects. So, like the steep steps of those pyramids in Teotihuacán, we took it up a notch, to do what we could to fix this disconnect.

Key Role of the Program and Portfolio Manager

By a *notch*, we mean that we have elevated our target to the individuals and organizations that manage *groups* or *collections* of projects (you'll learn more about this in Chapter 1 and in our "Project" dimension). We've moved up the pyramid of project, program, and portfolio management to the PMO level. Note that by *PMO*, we are referring to whatever entity in the enterprise oversees projects from a program and portfolio perspective. The names could vary considerably and could include PM Best Practices Office, Project and Program Management Office, or PM Center of Excellence. We are seeking out project managers as well as those who *oversee* project management as an operation or a discipline and are aiming to improve conditions not only for the project managers in their organizations but for better steady-state results and increased benefits realization.

Indeed, we've found that these ideas seem to get more traction, as we assert that projects, programs, and portfolios are the place where *strategy* (the action-oriented stepchild of mission, vision, and values) meets *operations* (see Figure I.3). While we have noticed that this gains more traction at the program and portfolio level, this book is certainly *also* aimed at the individual project manager, a change agent if ever there was one.

However, a funny thing happened as we thought about the pyramid and the idea of "rubber meets the road." We realized that we could express what we were saying, and even build a logical and productive assessment with a model that looked quite like a *wheel*. Talk about the rubber meeting the road!

Getting in "Shape"

Going back to Mexico for a moment, where did we head off to after our presentation? We were drawn inexplicably to the pyramids. And, as we climbed each (rectangular) step, we got closer and closer to our goal (the view from the top of the Pyramid of the Sun is outstanding). Stop for a moment, though. For who were the pyramids built? The Moon and the Sun. Spherical shapes. But project managers don't *like* circles and spheres.

Think about it. Project managers are constantly using triangles and pyramids.

For example, consider the following:

- Many PMs relate to the planning, construction (and ironically, the long-lastingness) of the Egyptian pyramid-building projects, often thinking of them, even idolizing them, as the predecessors of modern project management. We have the project–program–portfolio pyramid (Figure I.1):
- There's the triple constraint concept, which is often shown as a triangle or pyramid.
- And there's many more, some borrowed from other fields, such as the pyramidal Maslow's hierarchy, and the data–information–knowledge–wisdom pyramid.

And if it's not triangles we're drawing or referring to, we are also partial to rectangles, straight lines, arrows, and squares. We have risk maps, 2 × 2 matrices, network diagrams, inputs—tools and techniques—outputs, plenty of tables, and the list goes on and on. Then there's the Work Breakdown Structure, which we really like, because it sort of looks like a giant triangle made from rectangles, like the Pyramids of the Moon

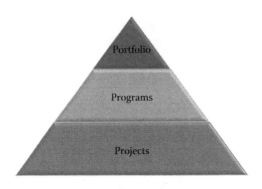

Figure I.1 The PPP pyramid.

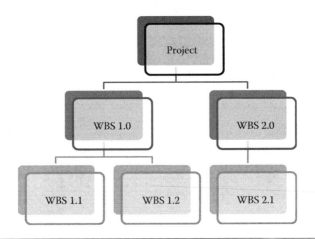

Figure I.2 Typical work breakdown structure: a pyramid built from rectangles.

and Sun, now that we think of it (Figure I.2). A triangle made from rectangles? That's project management *Nirvana*!

Not So Good at Circles

But circles…and spheres—we project managers rarely do well with them. Perhaps it's because they imply *endlessness*, where we crave *finality* in our projects. Perhaps it's because there are no corners in which to hide. In any case, they're a little uncomfortable for us project management types.

And perhaps that's just why a wheel (a sort of circle) is just *perfect* for conveying this message. We need a little "out-of-the-triangle" thinking!

A circle does indeed represent endlessness. This is the theme of Braungart and McDonough's *Cradle to Cradle—Remaking the Way We Make Things* (McDonough and Braungart 2009). It's used to describe situations where the end yields a beginning—the circle of life. And indeed, it has no corners—no places to hide. These are attributes we wanted from our model, attributes that would help transform project, program, and portfolio managers and to help them "get" the ideas that their business leadership colleagues, in increasing numbers, already "get."

We also chose a wheel to remind us that all of the necessary connection project, program, and portfolio managers have to bridge between strategy and operations. We *are* where the rubber hits the road—where ideas become reality—and in the case of this book, where a circle becomes a tire.

These are the wheels and cogs that mesh to drive things forward—to move ideas to reality (Figure I.3).

And we've chosen a wheel, because after all, the wheel can be looked at as one of the first inventions; it's a pragmatic, hands-on, real-world adaptation of a shape to serve a purpose and realize a benefit. The car

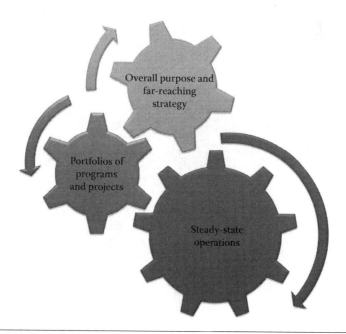

Figure I.3 Connection wheels/gears.

with circular wheels will win any race in which its rivals have triangular or square wheels, right?

Sustainability Wheel™

So, we've written this book for you "pyramid people," to help your enterprise succeed not only on individual projects, but in the world of programs and portfolios in the broader context of the enterprise, as well as the surrounding atmosphere in which the enterprise lives.

We hope that at a minimum, it strengthens your ability to make the connection—and we firmly believe it will help you transform that connection into a source of improvement for you, for your enterprise, the community, and the planet (Figure I.4).

The Sustainability Wheel identifies six interrelated dimensions of sustainability integration. It is composed of a *hub* (*Respect*), the sustainability philosophy of the organization, an adjacent ring (*Reflect*), which describes how that philosophy is conveyed internally, and the outer ring, containing the external facing dimensions. The outer ring (or tire) contains *Connect* (how external stakeholders view the organization), *Detect* (how the organization identifies, analyzes, and responds to sustainability threats, *Reject* (how the organization eliminates inefficiency and waste), and *Project* (how the organization identifies, develops, and measures their opportunities and converts them

Figure I.4   The sustainability wheel.

to projects and programs aligned with their central mission). Armed with valuable feedback on how an enterprise is doing in each of the dimensions, any business leader (consider that the project, program, or portfolio manager is a business leader) can determine which areas need improvement to help balance their sustainability efforts. The Sustainability Wheel can

- Evaluate existing sustainability programs and efforts
- Provide a baseline to measure present sustainability efforts
- Help determine priorities for the improvement of sustainability
- Validate that present sustainability efforts are within the organization's mission/vision
- Provide a mechanism to integrate sustainability into everyday operations
- Help convey the important role of project, program, and portfolio managers in the integration of sustainability at the strategic level of the enterprise

How to Use This Book

This book is organized to help you understand all of the organizational connections to sustainability, to determine where an organization is on a sustainability scale related to each of five dimensions: Respect, Connect, Detect, Reject, and Project, as well as to use the sixth dimension, the ability to Reflect, on the other dimensions. It can be used at all levels of the organization, but as we know, unless there is a commitment at the highest level of an organization, the message doesn't have the force it should. However, this tool is meant not only to be provided as a tool to measure an organization's connection, but as an easy-to-use artifact that can positively influence organizational leaders.

Chapter 1 is an important orientation for the reader. It provides the context of the intersection between projects, programs, and portfolios and the rationale and imperative to make that connection.

The context: Projects, Programs, Portfolios, and Strategic Implementation Management

- Findings from PMO Symposium
- Connection to Benefits Realization
- Gear Model

- PM Maturity (Bannerman Model)
- What is Success?
- Harry Mulisch *The Discovery of Heaven*—The beginning of the beginning, the end of the beginning, the beginning of the end, the end of the end.
- Project Success > Project Management Success (Drucker chart)
- 2 × 2 Matrices (Big Dig, Sydney Opera House, Edsel, Projects of the Year)
- Head, Hands, Heart
  - The wheel itself is change
- PMs like projects because they're *where the rubber hits the road*

The tool contains a set of questions designed to determine the measure of an organization's sustainability in each dimension.

The majority of the remainder of the book is the introduction to the Sustainability Wheel model and the assessment and coaching capability that it yields. As you read the chapters that make up the Sustainability Wheel (Chapters 2 through 4), consider the questions in those chapters at a high level, perhaps taking some notes in the margins about your own enterprise. We've written these "thinking questions" so that they provoke and focus your own considerations of the dimensions—and we've taken the next step as well, which is to explode or, to use a word we love as project managers, *decompose* them into a set of more concrete questions that you can use to actually derive an assessment of your enterprise along the six dimensions of the Sustainability Wheel.

The following is a list of the chapters, mainly devoted to the elements of the wheel. For illustration, we have also added the high-level question that is associated with each element. These will in turn yield the individual assessment questions.

- Chapter 2—The Hub: The Respect Dimension
  Have we clearly, accurately, and concisely stated our business case for sustainability in a way that is fully integrated into our raison d'être ? (or substitute mission/vision/values for the French phrase).
- Chapter 3—The Spokes: The Reflect Dimension
  How well have we conveyed our mission/vision/values to our workforce?

- Chapter 4—The Tire
  Connect
  What do *others* think of our CSR/sustainability efforts, especially relative to others in our industry or practice area?
  Detect
  How well do we identify, analyze, and respond to sustainability-oriented threats?
  Reject
  How well do we deal with process and product waste? Are we efficient in what we do?
  *Project**
  What is our level of project management maturity?
  How well do we identify, analyze, and respond to sustainability-oriented opportunities?

High-Level Questions from the Sustainability Wheel

Example high-level questions from the Sustainability Wheel—Chapters 2 through 4 (each category will have multiple questions, weighted to generate an overall score per category):

Respect: Does the organization's mission statement explicitly cover elements of social and ecological responsibility as well as financial responsibility?

Reflect: Do statistically significant polls of the organization taken at least annually indicate that employees are aware of and understand the triple-bottom-line nature of the organization's mission statement?

Connect: As measured by independent sources (such as the Newsweek Green Index or Climate Counts), does the organization consistently score in the top 20% of peers in its industry?

Detect: Does the organization explicitly list sustainability risks on standard risk register templates?

---

* *Note*: This particular dimension works off of two different meanings of the word project—project, as normally used in project management and, to project, a verb meaning to use a forward-thinking philosophy.

Reject: Is there a corporate Lean Six Sigma program in place?

Project: Does the organization have a Program Management Office (PMO) or Project Center of Excellence (COE) organization? Is it geared to identifying the possible opportunities available by considering all three elements of the triple bottom line?

Once the sustainability score for each dimension is determined, a "spider" chart can then be generated to visualize an organization's overall sustainability. Examples of the questions as well as examples of spider charts for a generally balanced organization (left) and one that needs to better balance their sustainability efforts (right) are included in the following. The spider charts were created by the authors for illustrative purposes only (Figure I.5).

Once the spider charts are generated and it is determined that some balancing of the organization's sustainability is suggested, the tool contains specific recommendations to move the organization toward a more balanced—and effective—sustainability effort. For example, the Bald Mountain organization results given earlier indicate a solid mission, a vision, and a set of values and that the company has conveyed this to the world; however, they have not yet engaged their employees—in particular, their project managers–to get traction in implementing sustainability in projects and operations.

The book provides specific coaching for various combinations of results. The "shape" of the radar chart is a signature of the organization's

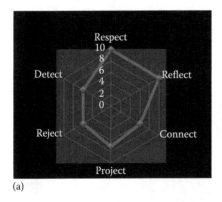
(a)

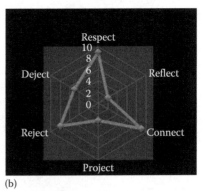
(b)

Figure I.5 Two example Sustainability Wheel™ signatures: (a) Interface global and (b) Bald Mountain.

sustainability behavior allowing directed feedback depending on that signature.

- Chapter 5—The Road
  This chapter is about putting the integration of sustainability into practice, fully engaging with your real-world stakeholders. In this chapter, we'll cover these aspects:
  Dialect: Setting up your enterprise for success by establishing a sustainability vocabulary and by expressing success in ways that powerfully convey your commitment to the long term, both inside and outside of your enterprise.
  Intellect: Benchmarking with other leaders, sharing common wisdom, understanding local, national, and global regulations—in general getting smarter about triple-bottom-line issues.
  Circumspect: The "feedback" loop of sustainability—in which we think about ways to continuously improve our performance. Reading this book and assessing your level of integration of sustainability into your portfolio is a big step forward here.

- Chapter 6—Interpreting the Sustainability Wheel
  Here, you'll find the condensed questions, instructions on how to approach the answers and scoring for your particular enterprise, and some sample results to show how to assess your results, including "signatures" of about 20 types among which you may recognize your own enterprise. Specific coaching is provided for these types.

References

Esty, D. and A. Winston (2009) *Green to Gold: How Smart Companies Use Environmental Strategy to Innovate, Create Value, and Build Competitive Advantage*, Wiley, New York.

Kiron, D., N. Kruschwitz, H. Rubel, M. Reeves, and S.K. Fuisz-Kehrbach (2013) *Sustainability's Next Frontier*, MIT/Sloan and Boston Consulting Group, Cambridge, MA.

McDonough, W. and M. Braungart (2009). *Cradle to Cradle: Remaking the Way We Make Things.* Vintage, London, U.K.

Senge, P.M. and B. Smith, and N. Kruschwitz (2010) *The Necessary Revolution: How Individuals and Organizations Are Working Together to Create a Sustainable World*, Crown Publishing, New York.

Authors

Rich Maltzman, PMP, has been an engineer since 1978 and a project management supervisor since 1988, including a two-year assignment in the Netherlands, in which he built a team of project managers overseeing deployments of telecom networks in Europe and the Middle East. His project work has been diverse, including projects such as the successful deployment of the entire video and telecom infrastructure for the 1996 Summer Olympic Games in Atlanta, and the 2006 integration of the program management offices (PMOs) of two large merging corporations. As a second, but intertwined career, Rich has also focused on consulting and teaching, having developed curricula and/or taught at

- Boston University
- Merrimack College
- Northern Essex Community College
- University of Massachusetts—Boston
- Clark University
- Benedictine University's Asia Institute

Rich has also professionally developed project management professional (PMP®) exam prep courseware, including exams and books. He even edited and was "the voice" for a set of eight audio CDs—a major part of a PMP prep course for an international company, for whom he has also facilitated PMP exam study groups. Rich was selected for the modeling team for the fourth edition of the *PMBOK® Guide* published by the Project Management Institute (PMI) and contributed to the chapters on quality and risk in both the fourth and fifth editions.

Rich has written and presented papers at international conferences of PMI and IPMA (International Project Management Association), and the Conference Board in South Africa, The Netherlands, Costa Rica, and Mexico City and at PMI Congresses in North America. He also presented, on the request of the Government of Malaysia, at their Green Technology conference in Kuala Lumpur in 2013.

Rich's educational background includes a BSEE from the University of Massachusetts, Amherst, and an MSIE from Purdue University. In addition, Rich has a mini-MBA from the University of Pennsylvania's Wharton School and a master's certificate in international business management granted jointly from Indiana University's Kelley School of Business and INSEAD of France. Rich received his PMP in 2000 and was certified by Change Catalysts as a CQ Certified Change Management Professional in 2015.

Rich has coauthored two other books—*Green Project Management* (CRC Press, Boca Raton, FL, 2010 Cleland Award Winner) with David Shirley, PMP, and *Project Workflow Management* with Dan Epstein (J Ross Publishing, Plantation, FL, 2014). Rich blogs regularly at earthpm.com and at projectmanagement.com.

Dave Shirley, PMP, has been an instructor and consultant, with more than 30 years of experience in management and project management, in the corporate, public, and small-business arenas. He is currently a graduate faculty member of Boston University, teaching courses in project management and developing and teaching Green IT. He developed, directed, and taught a project management–certification program at Northern Essex Community College in Haverhill, Massachusetts. He previously taught graduate courses in corporate social responsibility for Southern New Hampshire University Online and project management at New England College. Dave has presented papers on project management and sustainability at conferences in Costa Rica, Mexico, and the United States.

As a distinguished member of the technical staff with AT&T and Lucent Technologies Bell Laboratories, Dave was responsible for managing the first light-wave transmission products as well as several quality efforts. He was also AT&T's project manager for the first fiber-to-the-home effort in Connecticut and was the Lucent Technologies' program management director, managing several large telecommunications companies' equipment deployment.

Dave's educational background includes a BA in geology from Windham College, Putney, Vermont, and an honors MBA from Monmouth University in Long Branch, New Jersey. He also holds master's certificates in project management from Stevens Institute of Technology, Hoboken, New Jersey, and American University in Washington, DC, and is certified as a project management professional (PMP) by the Project Management Institute (PMI). Along with coauthor Rich Maltzman, Dave wrote the David I Cleland 2011 Award winning book, *Green Project Management*, CRC Press, 2011, and authored *Project Management for Healthcare*, CRC Press, 2011, and co-founded EarthPM (www.earthpm.com). He lives in Cape Neddick, Maine.

1
SUSTAINABILITY
AND SUCCESS

Begin with the end in mind.

–Steven Covey

We start our book by looking importantly at the *END* of projects. In fact, a subject of continual discussion on project management discussion groups is this: "What *is* project success?" When *is* a project's end? And we would add, when should the project manager's *view* of the project's *product* end? Indeed, should the project manager (PM) cast his or her view past the time when the project's product, service, or outcome has been handed over to the client, sponsor, or customer?

As attendees at the Project or Program Management Office (PMO) Symposium in 2013 in San Diego, CA, we were intrigued by comments from Project Management Institute (PMI) EO Mark Langley, in which he said that in his background as a member of the boards of directors of several large corporations, he never heard project management terms—*Gantt Chart, WBS*, and so forth—during board meetings. In fact, he rarely, if ever, heard the word "project" come up. The focus was on the operations of the companies—the deliverables in their steady state. Of course, we all know that it's the project and program managers, who enable these deliverables, but in order to communicate between the PM community and executives, it's important to have a common language—and it's not PM language.

This conference also reinforced our increasing view that the wwproject level and practitioners of project management were not necessarily the correct audience, or at least not the *only* audience for the message about sustainable thinking. So, many of the themes at the PMO Symposium were about strategy, mission, vision, and values, themes that were nearly absent at the PM conferences we've attended.

At that same conference, we were presenting about just this connection point. Our talk title was *Should your PMO serve as 'Chief Project Sustainability Office'?* and as you can tell from the title it was aimed at a broader view of project management—as is this book. Here, we "zoom out" from the project level to encompass programs, portfolios, and the complete enterprise-level view.

The Gear Model and Organizational PM

Since our book title includes the sequence "projects, programs, and portfolios," let's explain why we use those terms and how they're key to understanding the messages we want to convey. So, here's some important context. In fact, context will come up again later in this chapter (Figure 1.1).

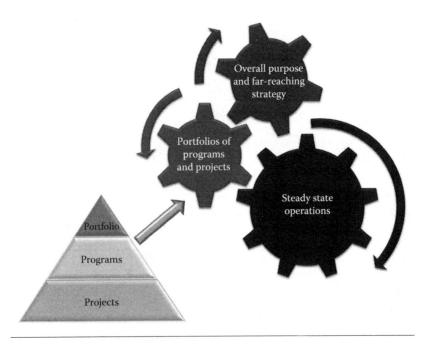

Figure 1.1   Pyramid and Gear Model together.

Organizational project management (OPM) is defined in the third edition of PMI's Organizational Project Management Maturity Model (OPM3®)* as

> a strategy execution framework that utilizes portfolio, program, and project management as well as organizational-enabling practices to consistently and predictably deliver organizational strategy to produce better performance, better results, and a sustainable competitive advantage.

As briefly mentioned in the "Introduction," the Gear Model is really our adaptation of Stanford's Strategic Execution Framework (SEF)[†] for a project management audience. In the SEF, the single point of contact between the "upper echelon" of an enterprise's structure—the mission, vision, value, its identity, structure, goals, and strategy—and its daily operations is through portfolios, programs, and projects.

Looking at these two models together, we can see that the connecting gear is what some may generally call "project management" or what we like to call "the art and science of project management." However, on closer inspection, this can be found to be made up of project management, program management, and portfolio management—in that hierarchical order.

Projects, Programs, and Portfolios

PMI now has standards for each of these levels as well as credentials to be gained by experience and knowledge at each of these levels. There are formal definitions of these levels which we summarize here, but before you get into the details, know that they represent—and are populated by—a large number of people who are focused by necessity on a shorter-term view. They are also of a "get it done" mind-set, which means that it's tough for them to conceive of a project being extended or—ironically—projected—into the distant future. So those readers who are sustainability focused, please appreciate that the other half of the readers who have a project, program, and/or portfolio mind-set

---

* Project Management Institute (2013) Organizational Project Management Maturity Model (OPM3®), 3rd Edn., Newtown Square, PA. http://marketplace.pmi.org/Pages/ProductDetail.aspx?GMProduct=00101463501.
† http://ipslearning.us/content/strategic-execution-framework.

may not understand that long-term view like you do. By the same token, nobody can get a group to focus on the work to be done and gain alignment and commitment like a project manager.

Projects are defined by PMI as "a temporary endeavor undertaken to create a unique product, service or result." *The PMBOK® Guide,* 5th Ed.* goes on to say that "the end is reached when the project's objectives have been achieved or when the project is terminated because its objectives will not or cannot be met, or when the need for the project no longer exists."

A *program* is defined as "a group of related projects, subprograms, and program activities managed in a coordinated way to obtain benefits not available from managing them separately." Importantly, this is the first time in the definition series that we see the word "benefits." We're getting closer to organizational goals. Note that the definition of the project does not have this thread. From the *Managing Change in Organizations: A Practice Guide,* "From a program point of view, the project outputs need to be assessed, not only on their intrinsic merit, but also on their contribution to … expected business benefits. Benefits can be measured only when the project deliverables are integrated into the operational process." Keep this in mind for an upcoming paragraph in this chapter in which we discuss the true definition of project success.

A *portfolio* refers to "projects, programs, sub-portfolios, and operations managed as a group to achieve strategic objectives."

Portfolio management oversees collections of projects and programs to assure that they are aligned with and integrated into the business so that they are realizing benefits—benefits that in turn are aligned with and in balance with the enterprise's mission/vision/values.

So now we definitively ascertained the connection to strategy and benefits.

In PMI's *Managing Change in Organizations: A Practice Guide,*† the connection is illustrated very well: "Organizational strategy is an input into the OP strategy execution framework and is based on the organization's mission, vision, and values. This strategy is developed

---

* Project Management Institute PMI (2013) *A Guide to the Project Management Body of Knowledge,* 5th Edn., Newtown Square, PA.
† Project Management Institute PMI (2013) *Managing Change in Organizations: A Practice Guide.* Newtown Square, PA.

to deliver maximum value to the organization's stakeholders and create the desired business results for the organization. This ... acts as the roadmap for the remaining elements of the OPM strategy execution framework."

Wow. Now we see the connection, even bringing in our theme of "the rubber and the road" metaphor (we like PMI's use of the word *roadmap*, they seem to be in alignment with us!). But notice that it's not until we step back and look at the three levels—project, program, and portfolio, and focus on OPM, that we see that projects don't just deliver any old "unique product, service or result," but rather very *specific* products, services, and results—ones that contribute benefits—sustainable benefits—to the organization's strategy, which in turn is (hopefully) connected to the organization's mission, vision, and values.

In our Gear Model, the Operations wheel does not turn without the connecting cog of project, program, and portfolio managers. Yet, because of the limited definition of a project to provide simply "a product," "a service," or "a result," project managers often work in somewhat of a vacuum, not knowing how their project is indeed part of a larger system, tied to the purpose of the enterprise. Yet, the coaching is there. Again from *Managing Change in Organizations: A Practice Guide*: "The organization uses program and project management as a means to effectively and efficiently deliver the initiatives of the organization. Strategy alignment continues through these disciplines and culminates with the realization of the value ... and is apparent when the results of the initiative are transitioned to the operations of the organization."

So, it's important to run projects well, but it's *imperative* to align them with the enterprise's organizational strategy. Figure 1.2—used with permission from PMI's *Pulse of the Profession* study of 2014, The High Cost of Low Performance*—shows that high-performing organizations (those with 80% or more of their projects meeting budget, scope, and timeline goals) are more than twice as likely as low-performing organizations (those with 60% or less of their projects

---

* Project Management Institute PMI (2014) *Pulse of the Profession: The High Cost of Low Performance*, http://www.pmi.org/~/media/PDF/Business-Solutions/PMI_Pulse_2014.ashx (accessed September 14, 2014).

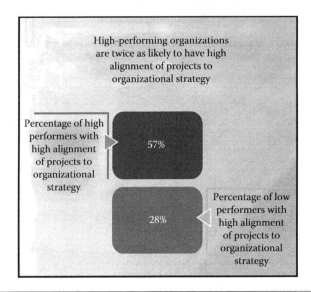

Figure 1.2 PMI statistics from *Pulse of the Profession*. (From PMI, *Pulse of the Profession: The High Cost of Low Performance*, http://www.pmi.org/~/media/PDF/Business-Solutions/PMI_Pulse_2014.ashx, 2014.)

meeting budget, scope, and timeline goals) to have high alignment of projects to organizational strategy. So, there are significant study data to show that this model represents reality.

The Need to Integrate—Not Add

In their paper "What can project management learn from considering sustainability principles" (Gareis, Huemann, Martinuzzi),* they identify that "In many project-oriented companies, who claim to consider sustainability, it remains to be the responsibility of the Sustainability Office and is not build into the business processes of the company." In particular, they say, "Most companies do not consider sustainability principles in their projects, although the management of projects makes an important and significant contribution to value creation globally." Why does this happen? It's partially because of a focus on the near term—only the near term. As Gaeis et al. say,

---

* IPMA (2011) *Project Perspectives*, R. Gareis, M. Huemann, and R.-A. Martinuzzi (2011), What can project management learn from considering sustainability principles?, http://ipma.ch/assets/re-perspectives_2011.pdf.

"The sustainability of ecosystems over time as well as the consideration of the needs of future generations are in direct contradiction with today's ever shortening time horizon of decision–makers," and with respect to projects, they say, "The temporary character of projects contradicts the long-term orientation of sustainability." This chapter goes on to show how enterprises can—and should—reconsider their orientation, at least their project orientation, to integrate sustainability principles not only for the altruistic rationale of doing so, but because this integration comes "back around" to provide benefits even in the short term, such as improved innovation, reduced waste, higher morale, and simply better-executed projects that are more fully in line with the enterprise goals and objectives.

NASA is an example of a highly (pun intended) visible, and highly project-based organization that has made this connection and which is serious about integrating sustainability into the enterprise—and in particular, project management.

At a 2011 presentation in Dallas, TX, given at PMI's *Sustainability and Project Management: The Future is Now*, Research Working Session, Olga Dominguez of NASA describes the motivation for and implementation of such an integration of sustainability into its culture (Figure 1.3).

Figure 1.3 NASA integrates sustainability.

Integrating Sustainability: A Key Trend

ESI International publishes their top trends in project management* each year in early January. For 2015, their top 10 list of trends includes not 1, but 2 trends that are directly "on message" with regard to integration of sustainability and project management. Here, they are—their #5 and #9:

*#5: Organizations Must Build Bottom-Up Processes to
Link Project Outcomes to Organizational Strategy*

When we see "bottom-up" here, we can go back to our Gear Model. The "bottom" here is operations—the steady-state result of projects, programs, and portfolios. And those projects, programs, and portfolios represent the connecting gear that allows the enterprise's purpose (think mission, vision, and values) to turn the "producing wheel" for stakeholders. However, that only happens if the project manager looks "up" and "down" to know the purpose of the company and the longer-term effects his or her project's product actually has.

*#9: Project Management and Business Strategy Better
Align to the Benefit of the Organization*

Alignment of projects, strategy, and steady-state benefits—this is the main topic of our book. It's similar to what we covered in #5, except this time the focus is on "benefits." When we see the word "benefits" we think of success. Later in this chapter, you'll find a full section on success. For now, think of it this way: there is a huge difference between *Project Management Success* and *Project Success*. The former is about meeting deadlines, sticking to a schedule, staying under budget, and providing deliverables. Don't get us wrong; these are all worthy, important, difficult things to do, which require PM sophistication and excellence. But in the scheme of things, it's extremely narrow in its vision and viewpoint. *Project Success*, on the other hand, is

---

* http://www.esi-intl.co.uk/blogs/pmoperspectives/index.php/10-project-management-trends-watch-2015/.

focused on the long-term benefit realization provided by the project's product. It considers the economic, ecological, and social by-products of the product while it's operating for 1 year, 2 years, 10 years, and 500 years. It's holistic. And a focus on *Project Success* yields better project management because it assures that the project is aligned with the enterprise's purpose, which inevitably is geared around longer-term, triple bottom line (TBL) concerns. To put it in concrete terms, where *Project Management Success* might be represented by a Cost Performance Index (CPI) of 1.09, *Project Success* would be represented by a *product* that responsibly delivers a profit for the organization for 5 years.

An Example of Integration: The Sustainability Breakdown Structure

One way to integrate sustainability into projects—and more importantly into the project managers' mind-sets, is to use tools already familiar to project managers in ways that bring sustainability thinking into the forefront. This can be as simple as assuring that risk identification includes TBL considerations, adding these elements to risk register templates, or it can be adapting an entirely new idea into a common format. Given the propensity for project managers to think of the world in terms of "breakdown structures" such as the Work Breakdown Structure or Organizational Breakdown Structure (see our "Introduction"), we think project managers will quickly line up in support of a Sustainability Breakdown Structure. The idea here is to put the project's end result at the top and use the TBL elements: Economic, Ecological, and Social as the first level of decomposition. Then, with the long term in mind, further decompose these branches. Unlike a Work Breakdown Structure (WBS), where the project manager seeks to determine schedulable, assignable "work packages," here the PM focuses instead on "use and disposal packages"—contributions of the project's product to the enterprise's portfolio and strategic goals, as well as impact (both positive and negative) on the surrounding environment including the ecological and social systems that this product will affect in its steady-state use and even in its disposal. This forces the project manager to think long term, but in a way that is familiar. In the following, we've started a simple example using a bridge project (Figure 1.4).

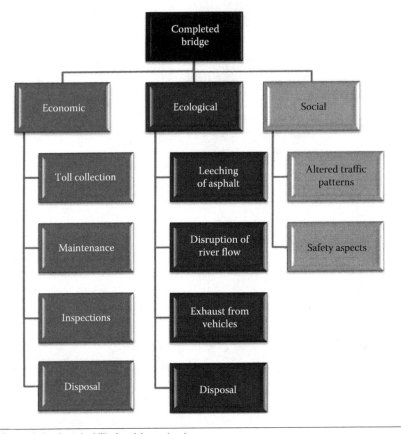

Figure 1.4 Sustainability breakdown structure.

Creating Shared Value

At the time our last book, *Green Project Management,** was published, we saw more and more about sustainability being adopted not as an "add-in option" for business, but as an integrated approach to improve results—to realize benefits.

One example of this is famed author and marketing guru Michael Porter and his concept of Creating Shared Value (CSV).

CSV is about seeing the opportunities, not the costs, of addressing social and environmental issues. The "kneejerk reaction, that any social issue or dealing with any social problem actually creates cost for the company and reduces profitability, that whole instinct is flawed,"

---

* R. Maltzman and D. Shirley (2010) *Green Project Management*, CRC Press, New York.

he says. "The more you start to help companies see that this is really about productivity and efficiency and the use of technology..., then all of a sudden the whole level of energy, focus and the results dramatically improve."

Porter began looking into environmental issues while conducting research for his book, *The Competitive Advantage of Nations*.* Among many other insights, Porter found that the countries that implemented strong environmental regulations were actually more competitive than those that did not. This was a very different finding from the thinking of the time.

Then, in 2006, Porter and fellow Harvard Business School colleague Mark Kramer introduced the concept of CSV in a *Harvard Business Review (HBR)* article[†] and the pair expounded the concept in a follow-up *HBR* article[‡] in 2011. At the heart of the CSV framework is the idea that the competitiveness of a company and the systems surrounding it are inter-reliant. And if companies act on this interdependency, they can create value for their businesses as well as society. In turn, this could launch "the next wave of global growth."

Indeed, companies like Nestle have adopted this as part of their publicly stated business strategy. Directly from Nestle's website:

> Creating Shared Value begins with the understanding that for our business to prosper over the long term, the communities we serve must also prosper. It explains how businesses can create competitive advantage, which in turn will deliver better returns for shareholders, through actions that substantially address a social or environmental challenge. As a company, we are best positioned to create shared value in three areas: nutrition, water and rural development.

Nestle has even created a page (http://www.nestle.com/csv/what-is-csv/commitments) where its measurements of progress against CSV commitments can be seen.

---

* M.E. Porter (March–April 1990) The competitive advantage of nations. *Harvard Business Review* 68(2): 73–93.
† M.E. Porter and M.R. Kramer (December 2006) Strategy and society: The link between competitive advantage and corporate social responsibility. *Harvard Business Review* 84(12): 78–92.
‡ M.E. Porter and M.R. Kramer (January–February 2011) Creating shared value. *Harvard Business Review* 89(1–2), p. 1.

Table 1.1   Nestle's CSV Paradigm

COMMITMENT	HOW	PROGRESS	OBJECTIVE
Provide nutritionally sound products designed for children	Nestlé Nutritional Profiling System/ Nestlé Nutritional Foundation criteria Nestlé Children's Healthy Growth Strategy	At the end of 2013, 96% of our products met all of the Nestlé Nutritional Foundation criteria for children (2012: 89%), which are based on international public health recommendations, such as those of the World Health Organization and the Institute of Medicine.	*By 2014*—100% of our children's products will meet all Nestlé Nutritional Foundation criteria for children.

For example, in the area of nutrition, there is a commitment for providing nutritionally sound products for children. Here is how it appears in a CSV paradigm (Table 1.1):

Benefits Realization and the Real Definition of "Project Success"

With the model of projects, programs, and portfolios in hand, along with the knowledge that these comprise the wheel connecting an enterprise's vision to its operations, we can see that a project, program, and/or portfolio manager is producing a better product, service, or outcome if they know the context—the other two gears in this case. Also, we know from thinkers like Porter and from studies by PMI, that there are compelling reasons to realize the benefits of sustainability thinking. And that leads us to a discussion about benefits realization and project success.

As Baker, Murphy, and Fisher said decades ago, in their 1983 book *Project Management Handbook: Factors Affecting Project Success*,* there is no such thing as absolute success in project management—only its "perceived success." They also said that the ways in which a project's success is evaluated probably changes over time. Keep this thought in mind through the remainder of this section.

One doesn't have to go back to the early eighties to find this concept. In the December 2014 issue of *PM Network* magazine, Gary

---

* N. Baker, D. Murphy, and D. Fisher (1983) *Project Management Handbook: Factors Affecting Project Success*, Van Nostrand Reinhold, New York.

Heerkens, PMP, in an article called "Generating Value," opens the article with this:

> Which is better: a project that runs on time and budget and meets its objective, or a project that runs late and is over budget but overachieves on cost savings that have a "change for life" impact?

Heerkens goes on to describe survey responses to this question, in which only about 25% "answered the question in appropriate, business-based manner," meaning that they opted for an economically sustainable version of success over *project management* success. Heerkens, noting that about 75% failed to "elevate their perspective beyond the realm of the triple constraint," says, "this does not bode well for project managers seeking to function in a project environment that is increasingly business focused." Note that this only includes economic sustainability, and if it had included ecological and social aspects, it would probably have revealed a much higher percentage of those project managers who failed to elevate their perspective to a broader, longer-term view.

Significant research* (Ika, 2009; MacLeod et al., 2012) into the subject of success in projects illustrates the increased focus and complexity of this subject. We sum it up with a simple equation as given in Figure 1.5.

In words, this just says, "over time, project success is much more than just project management success."

Although it is critical to manage projects correctly (i.e., "Project Management Success")—this means good use of standard PM tools, great communications among the project team, all of the things that outstanding project managers do to get terrific project results, like being on time, under budget, and within scope. Yet, while it's

$$\sum_{t=0}^{\infty} \text{Project Success} \geq \text{Project Management Success}$$

Figure 1.5 Our equation for project success.

---

* I.A. Lavagnon (2009) *Project Success as a Topic in Project Management Journals* 40(4): 6–19.

L. McLeod, B. Doolin, and S.G. MacDonell (2012) A perspective-based understanding of project success. *Project Management Journal* 43(5): 68–86.

Table 1.2 Drucker vs. PM View

Drucker's view of success	*Efficiency*: "Do Things Right"—the process of Project Management
	Effectiveness: "Do the Right Things"—connect your project to strategies and business objectives of your organizations
Sustainable view of success	*Endurance*: "Do the Right Things Right With Lasting Power." In other words, be efficient and effective with the triple bottom line in constant focus.

obviously critical to accomplish those projects' objectives, it is also important to look at the projects' *outcomes* in the *longer term* and more *holistically* (i.e., "Project Success"). This not only includes market impact, competitive impact, investor impact, and industry impact, but also, we assert, includes the temporal aspects—the lasting impacts—ecological, societal, and economic.

Let's clarify this equation further by using Peter Drucker's definition of success* using efficiency and effectiveness, but adding the element of "Endurance" to his use of efficiency and effectiveness (Table 1.2).

An interesting corollary to this idea—and one that is a graphic reminder of its importance—is that projects and their outcomes can be assessed in radically different ways. We've provided our own illustration of this (see Figure 1.6). For example, the Sydney Opera House was considered a failure from a project management perspective. After all, it was planned to be a U.S. $7M, 5-year project and instead cost U.S. $110M and took 13 years to complete. However, its outcome—the Opera House *itself*—is considered a national treasure and a tremendous place to take in any event.

In the opposite corner, we have projects that may have been within budget on schedule and even producing the requested project scope, but the outcome's (the project's product's) long-term value was very low or even negative to the organization and its customers. The lower left quadrant represents projects (like Boston's Big Dig) that were over budget, late, and didn't deliver proper scope—and has ongoing operational problems (water leaks, falling ceiling tiles, and corroding light fixtures). The upper right quadrant is the area to which we aspire—a project that does the right things right. And to this, we add the aspiration to do the right things right with the TBL in mind.

---

* P.F. Drucker (1986) *Managing for Results*, 2nd Ed., HarperBusiness, New York.

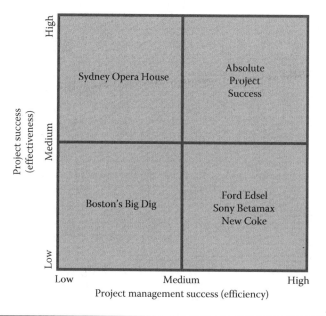

Figure 1.6 Project vs. project management success.

In this view, a project manager's time frame is necessarily expanded into the future. In fact, we imagine a third dimension to this chart (Figure 1.7), adding the enduring success—and TBL impact of the project's product. We do not assert that the PM must be held accountable for the long-term, steady-state operation of their project's product. However, we do firmly believe that the PM sacrifices valuable input for their project management success (efficiency) and the project success (effectiveness) if they do not take that long-term view while initiating, planning, executing, monitoring and controlling, and closing their projects.

Project Success and Improved PM Maturity

In the Spectrum of Green from *Green Project Management*, we can see a range of projects that go from those with no apparently "sustainability" connection to those which are by nature geared to sustainable outcomes. This is useful to determine the expectations for the project manager's needed focus on sustainability based on that attribute.

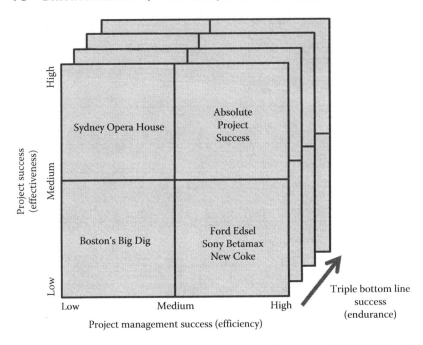

Figure 1.7 View of project success—With triple bottom line.

A broader framework to describe the intersection of project management, project success, and sustainability is the Multilevel Project Success Framework* (Bannerman, 2008). In this view (see Figure 1.8), success can be seen from left to right with an increasingly expanded, holistically viewed and long-lasting way. One could almost look at this as a maturity model in which the left side is a narrowly focused project organization. Movement to the right indicates benefits realization maturity.

This model does not mention sustainability by name but clearly *benefits realization, unintended impacts, market impact,* and *connection to goals/objectives and the business plan* are all concepts aligned with our view of integrating sustainability into project management.

In many cases, project experts are exploring the elements of project success without using sustainability by name, but clearly covering the right territory. The following example is used with permission

---

* P.L. Bannerman (2008) Defining project success: A multilevel framework. *Project Management Institute Research Conference*, Warsaw, Poland, pp. 5–6.

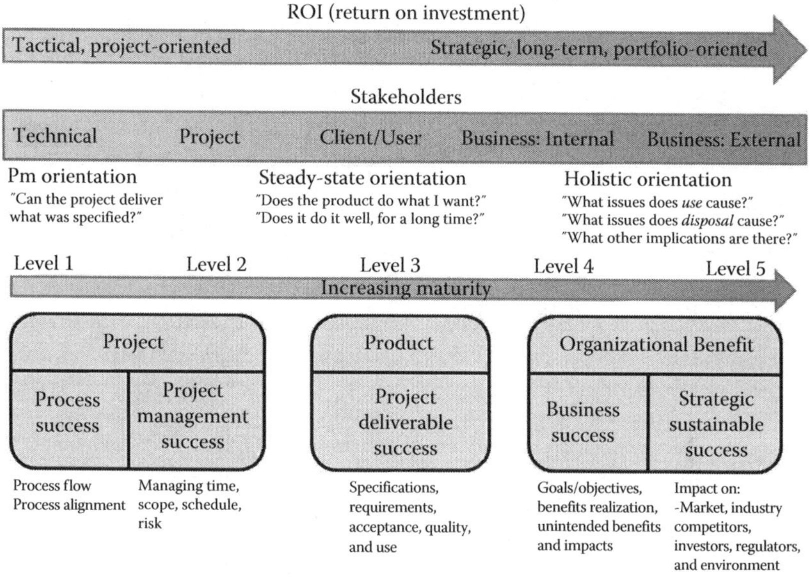

Figure 1.8 Multilevel framework. (Adapted from Bannerman, P.L., Defining project success: A multilevel framework, *Project Management Institute Research Conference*, Warsaw, Poland, pp. 5–6, 2008. With permission.)

from the blog of Success-Unique-Knowledge-Attitude-Development (SUKAD), authored by Mounir Ajam.*

Almost two years back, we were in such a class. The client was an organization that sells valves and pipes for the petroleum industry. The participants were from different functions. A group of the participants wanted to work on a project, to build a gymnasium at one of its warehouses (facility); this was a real project they were considering.

Based on the previous paragraph, what is the project?
Most will say, build the gymnasium.
Is it?
Is the company in the business of building gymnasiums?
No!
OK, then what is the project?

Maybe we should ask the question differently, what is the business objective for the project, the business driver; business case?

---

* Mounir Ajam. SUKAD. http://blog.sukad.com/.

Uh!

Now we understand.

In this case, we can say "improve employee health."

Consequently, the project is to build a gymnasium to improve employee health.

Great – now we know the project.

Let us pose for a minute; did you realize how one statement, the project idea statement, could mean totally different things? If the team does not understand the business objective, then how can we deliver a successful project?

As you can see in Figure 1.9, there are different views of success—different *levels* in SUKAD's model.

We also acknowledge that project success has different levels, and in line with the model given earlier we see them in terms of their willingness to take a longer view of the product of the project. To illustrate our view, let's take a look at the question, "when is the end of the project?"—and to make it interesting, we use the table of contents of an unlikely but famous Dutch novel as our guide.

As project managers, we think (and often dream) about the *end* of our projects. And *what is that end?*

It's the successful turn up of a computer network. Or, it's the availability of a new service. Or, perhaps it's the readiness of a new bridge, a new building, a new drug, and a new electric toothbrush.

Do you note a pattern here?

When we're *done*, something *else*, usually something *bigger*, is *starting.*

We often limit our thinking as project managers to the life cycle of our project and *don't think enough about the life cycle of the product of our projects.* For those of you who weren't paying attention, that's the "bigger thing" that is enabled by your project.

This concept reminded me (or perhaps it's the other way around) of a book and movie which I've always enjoyed. Both the book and movie share the title, *The Discovery of Heaven.** Do yourself a favor and rent this film and/or read the book. It's by famed Dutch author

---

* H. Mulisch (1998). *The Discovery of Heaven*, Penguin Books, London.

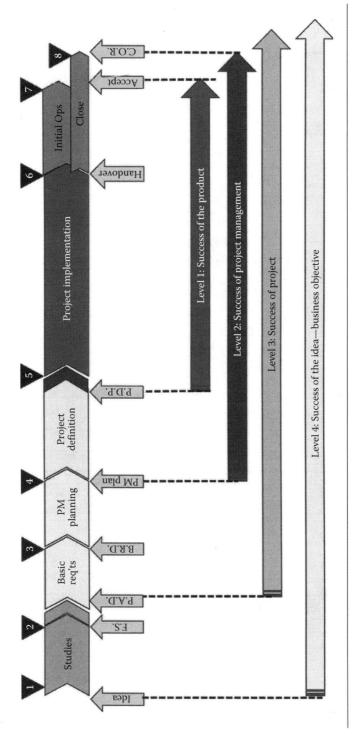

Figure 1.9 Varying views of success—SUKAD model.

Harry Mulisch, and it's not known so well in the United States but is throughout Europe. From Wikipedia, here is the synopsis of the book:

> The Discovery of Heaven tells the story of an angel-like being, who is ordered to return to Heaven the stone tablets containing the ten commandments, given to Moses by God, which symbolize in the book the link between Heaven and Earth. The divine being, however, cannot himself travel to Earth, and on several occasions in the book resorts to influencing events. He affects the personal lives of three people (two men and one woman) in order that a child will be conceived. This child would then have an innate desire to seek out and return the Tablets.

The inspiration for the posting comes not from the religious theme of the book, but rather the way the book is organized as follows:

- The Beginning of the Beginning
- The End of the Beginning
- The Beginning of the End
- The End of the End

Now back to project management and our way of thinking. We don't realize it when we plan our projects but *we only work on the equivalent of the first two parts of Mulisch's book*: The Beginning of the Beginning, where we Initiate and Plan the project, and The End of the Beginning, in which we execute, monitor and control, and close the project. We don't look ahead—often enough—to what happens in the "life and death" of the bridge, the building, the drug, the computer system—even the electric toothbrush.

Strategy, Projects, Programs, Portfolios, and Success

This longer-term view touted by ourselves, by Bannerman, and by Ajam is important not only for the ecological and social aspects of the TBL, but is also essential for the economic sustainability of the enterprise and the proper execution of an enterprise's strategy.

The PMO is often the organization that can help provide this view.

In 2013, the PMO Symposium in San Diego, CA, featured the connection between the strategic initiatives of an enterprise and the delivery of projects. Attendees received a large packet of research focused on this topic. A significant portion of the published research focused on this idea and the key role that projects, programs, and portfolios played in delivering business results and the often weak connection between a company's mission, vision, values and strategy, and the day-to-day operations. In a 2013 PMI Pulse of the Profession report,* "The Impact of PMOs on Strategy Implementation," the overriding theme is that "the alignment of the PMO to the goals of the organization is the key to driving strategy implementation." High-performing PMOs—those that achieve 80% or more projects on time, on budget and meeting original goals—are more than three times as likely as lower-performing organizations to reach their full potential in *contributing business value* to their organizations. The conclusion of this report is that "Alignment of projects to the goals of the organization is, therefore, instrumental in increasing business value. PMOs that are frequently involved in project alignment to strategic objectives of the organization are nearly twice as likely to be a high performing rarely involved." So, it pays for an enterprise to assure that their projects are aligned to strategy, mission, vision, and values. The remainder of this book continues to illustrate that point and importantly, to show how this can be done for all three elements of the TBL, which often actually reinforce each other.

In another report from the packet, "Why Good Strategies Fail—Lessons for the C-Suite," published by *The Economist*† discusses closing the loop of strategy formulation and implementation. We would say that this is another way of saying "making sure that the rubber hits the road." In this report, Jeff Austin, a DuPont Pioneer, is quoted as saying, "Ensuring a tight linkage between strategy development and how that translates operationally is a challenge. But in effective companies they are integrated in a holistic way ... people should see these steps as part of a continuum." That continuum is the connectivity between mission, vision, values, strategy, portfolios, programs, projects, and operations. We would add that the wheel (or gear) that connects strategy to operations—the portfolio of programs and

---

* http://www.pmi.org/~/media/PDF/Publications/PMI-Pulse-Impact-of-PMOs-on-Strategy-Implementation.ashx.
† http://www.economistinsights.com/analysis/why-good-strategies-fail.

projects—has got to be engaged to the driveshaft (strategy and vision) and the road (operations) for the vehicle to be able to move. That's the Sustainability Wheel.

An additional report in the PMO Symposium package, *Strategic Initiative: Management: the PMO Imperative**—published jointly by PMI and the Boston Consulting Group, summarizes the ways in which PMOs can evolve toward a tighter connection between strategy and operation. The four imperatives are as follows:

- Focus on critical initiatives
- Institute smart and simple processes
- Foster talent and capabilities
- Encourage a culture of change

We see all of these through the lens of sustainability and TBL thinking.

For example, the assertion, "Focus on Critical Initiatives," to us, includes the statements and obligations an enterprise makes in terms of not only short-term financial goals, but longer-term economics, social responsibility, and commitment to the environment. This means that if an enterprise states on the "About Us" section of its web page that it is committed to reducing waste and emissions, this must be "ever present" on the minds of functional managers in the organization but importantly *also* on the minds—and *charters*—of project managers in the organization. We see organizations that include TBL elements in their annual reports as truly buying into this concept.

"Instituting smart and simple processes" evokes the "SMARTER" principle from our earlier book *Green Project Management* which adds "Environmentally Responsible" to the normal "SMART" objectives of "Specific, Measurable, Attainable, Realistic and Time-bound." It also can refer to the idea of lean processes and the application of life cycle assessment to root out unneeded effort and waste in processes.

"Foster talent and capabilities," to us includes the elements of training project managers to—while retaining their necessary focus

---

* https://www.pmi.org/~/media/PDF/Publications/BCG-Strategic-Initiative-Management.ashx.

on the milestones and shorter-term goals of their projects—retain a broader perspective in terms of time frame and scope. It includes adding to project managers the capability to think more strategically and more connectedly to the mission, vision, and values of their enterprise.

And finally, "Encourage a culture of change" echoes our sentiment expressed in Green Project Management, that project managers are, after all, agents of change, and thus become an ideal candidate to implement sustainability thinking in their organization. It doesn't have to be a Ray Anderson, a CEO with an epiphany that triggers change. Why not the Project Manager?

Table 1.3 summarizes these initiatives in a view that shows a "necessary evolution" from the current state to a future state in which PMOs have adapted these imperatives.

Finally, the paper, Strategic PMOs Play a Vital Role In Driving Business Outcomes,* a thought leadership paper commissioned by PMI, and written by Forrester research, does an outstanding job of conveying how the PMO (and thus Projects, Programs and Portfolios) contribute to connecting the mission, vision, and values to "the road."

Table 1.3   The Role of the PMO in Connecting the Rubber to the Road

PMO IMPERATIVE	CURRENT STATE	FUTURE STATE
	PMO focus is on tracking and reporting on project and program processes	PMO focus is on actively supporting delivery of fullest program value with minimum sufficiency
Focus on critical initiatives	Providing too little or too much information, and focus is on activity completion	Providing meaningful, forward-looking information tied to delivery of strategic impact
Institute smart and simple processes	Focus is on process and policing the process	Focus is on minimum sufficiency in processes in order to generate progress
Foster talent and capabilities	Fostering project and program management skills	Fostering the Talent Triangle: project and program management skills, business acumen, and leadership
Encourage a culture of change	Limiting alignment and connection to strategy with accountability limited to project metrics	Establishing clear alignment with the senior leaders accountable for change

* http://www.pmi.org/~/media/PDF/Publications/Forrester-PMOs-Play-Vital-Role-TLP-PMI-Final.ashx.

Forrester conducted in-depth interviews with 40 leaders of PMOs. In the executive summary of this paper, they say, "it was clear that PMOs strategically aligned with executive management played a direct part in enabling their companies to obtain successful business outcomes." Not *project* outcomes. *Business* outcomes. We're talking sustainable success here, not project success. "By investing in a strategically aligned PMO, every company saw distinct benefits; two-thirds of the companies interviewed saw improved performance in less than six months and realized the value of investing in the PMO within two years." Key findings of this study were that successful PMOs

- Have a seat at the executive table
- Are a vital part of the strategic planning team
- Embrace core competencies
- Use consistent objectives across industry and regions

These attributes are in line with our assertion that anything which consistently and fundamentally connects projects and the oversight of projects (which really is what a PMO is about) with the strategy of an organization is going to yield benefits. In addition to this, it's increasingly common to find that enterprise strategy now includes TBL thinking, and you can see that the PMO needs to be a vehicle to assure that the enterprise's projects are including TBL thinking. In fact, the very presentation we gave at this PMO Symposium was as follows: "Should your PMO serve as Chief Project Sustainability Office?"—asserting the need for PMOs to do precisely what this research indicates *needs to happen*.

You've Had the Power All Along—and Our "3-Click Challenge"

In the classic 1939 movie, *The Wizard of Oz*, the *Good Witch of the North*, Glinda, comes to Dorothy and advises her that she had the power all along to go home. All she had to do was "tap her heels together three times" (of course she had to be wearing the proper footwear—ruby slippers) and she would be brought home "in two seconds."

What could this possibly have to do with project management and/or sustainability?

A lot! And we think this is very worth your time if you'll follow us on this short journey.

Let's start with change.

Dorothy wanted to *change* her location. The lion was nice, as was the scarecrow and the tin man. Still, she was homesick. After all, she belongs in Kansas, "in her own back yard," as she says herself. She wants this change but she believes that she will need to call on much higher powers (witches, wizards) to do so.

But Glinda tells her that she just needs to *tap her heels thrice* to go home. Click. Click. Click. Three simple clicks. That's all it takes.

Now, here's where we'll make the connection. As authors of the book *Green Project Management*, we've gone around the world to talk about how project managers need to bring sustainability thinking into PM, we get lots of pushback. Some of it is due to folks just not believing in the concept of sustainability in the first place—or at least in the ideas of climate change, often politicizing the issue and making it one of left or right.

Forget that.

We're not talking politics here.

In our collection of "pushbacks" we also get folks who (luckily) get past the politics of all of this, and realize this is a real issue but tell us things like this:

- I can't make a decision to purchase a more sustainable raw material for my project if it costs more.
- I don't have the authority to bring sustainability into a project, never mind into my company's thinking.
- I'm just a project manager...

These folks are acting just like Dorothy. Guess what, project managers? Don't undersell yourself! Don't be a "Dorothy." You *are* change agents. By *definition* you are change agents. Projects, by definition, are about change. *Nobody would do a project in the first place if they didn't want something to change.* You are wearing ruby slippers! You had the power all along! And not only that, your enterprises are likely already promoting sustainability at the highest level, so by being a change agent and making decisions (for example) to use a slightly more expensive vendor or material because it has a more sustainable long-term result, you are actually acting in line with your top management's strategic objectives.

Now we know that real life isn't exactly like Oz. It's probably not that straightforward. You may have to get some other goblins and flying monkeys (usually known in our world as middle managers) out of the way, but up at the higher echelons of your company, they're rooting for YOU to be the change.

And there's the other connection to the "three taps." In this case, we've done a little homework and we're asking you to do the same in your enterprise.

We assert *that within three clicks on your company's EXTERNAL home page, you will end up on a page devoted to either sustainability, triple-bottom-line thinking, corporate social responsibility (CSR), or something of that ilk.* And that page will have statements, objectives, goals, values about how your leadership views itself, views its responsibilities to the TBL, and by extension, how it views you—Mr. or Ms. Project Manager—Mr. or Ms. Change Agent—as a way to get their vision to steady-state operation.

We've tried it. It works.

Here are just a few examples of major global companies and within three clicks, we had found our way to Kansas, or at least to the sustainability pages of these companies. We bet that it will work for yours as well.

- Nike
- Patagonia
- BP
- McDonald's
- Microsoft

Making the Change to Sustainability Thinking in Projects, Programs, and Portfolios

If we stick to the Wizard of Oz theme for one more cycle, we know that Dorothy was accompanied by three friends during her journey to the Emerald City. They were the Scarecrow, the Tin Man, and the Lion. If you'll allow us a bit of artistic creativity, you can imagine that the scarecrow was the brains, the tin man was the heart, and the lion (for the sake of this discussion) was the paws (or "hands") of their journey. Keep this in mind as we discuss head, heart, and hands in terms of change management.

Table 1.4   Strengths and Opportunities for Change Styles

	LEADING CHANGE FROM THE HEART	LEADING CHANGE FROM THE HEAD	LEADING CHANGE FROM THE HANDS
Change leader style defined	Engaging, caring, people-oriented change leader	Strategic, futuristic, purpose-oriented change leader	Efficient, tactical, process-oriented change leader
Strengths	Motivating and supportive coach	Inspirational and big picture visionary	Planful and systematic executer
Developmental opportunities	May neglect to revisit overall change goals and not devote attention to the specific tactics of the change process	May "leave others behind" wanting to move sooner than people are ready and lack detailed planning and follow-through	May lose sight of the "big picture" and devalue team dynamics and individual's emotions

One of the best books on change management as applied to project management is *Change Intelligence: Using the Power of CQ to Lead Change That Sticks** by Barbara Trautlein.

In the book, Dr. Trautlein walks us through her model of change management, which asserts that people lead change from the heart, head, and hands—each of us has a propensity to lead the change from one—or a combination —of these tendencies.

In Table 1.4, you can see the three change styles summarized:

Dr. Trautlein goes on to discuss in particular how an understanding of one's own tendencies can help one lead change in various contexts, including that of project management. As she says, "we know that executives initiate change and supervisors implement them. Similar to executives, PMs can influence an initiative's overall direction, but typically they're not yet strategic leaders. Similar to supervisors, PMs are accountable for executing change, but they have to operate on a more tactical (hands-on?) level as they plan and coordinate … a process that typically involves people from multiple departments."

Trautlein goes on to discuss how project managers lead change from the heart, head, and hands; see Table 1.5 adapted from her book:

The Change Intelligence (CQ) model includes an assessment (available with purchase of the book) which can tell you where you "reside"

---

* B. Trautlein (2013). *Change Intelligence: Use the Power of CQ to Lead Change That Sticks*. Greenleaf Book Group Press, Austin, TX.

Table 1.5   How a PM Uses Heart, Head, Hands

As a project manager, you…

HEART (AFFECTIVE)

Engage and motivate others—who may only report to you part time, and over whom you may have
  no formal authority

Build a cross-functional team—across disciplines you may not be expert in yourself

HEAD (COGNITIVE)

Communicate goals—not just to your project team members but also to their managers,
  departments, down and up

Coach others through giving and receiving feedback; ensure they have the training they need to be
  effective project team members

HANDS (BEHAVIOR)

Build, own, and manage the plan relentlessly; resolve issues and mitigate risks through your
  ability to influence versus mandate

Provide resources—often through negotiation with other groups across and up the organization

in the combination of the three scales (heart, head, and hands). The
three elements are assessed and yield seven different regions or styles
(Figure 1.10).

As a project manager, it's likely that you'll end up as driver, executer,
or facilitator, the "high hands" styles. The book provides coaching for

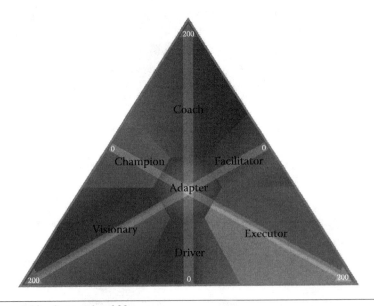

Figure 1.10   Seven styles of CQ.

individuals in all seven areas, here, for our PM colleagues, we give some of the top tips for the "high hands" styles.

The overarching comment for high-handers (see Table 1.4) is this: you may lose sight of the big picture and devalue team dynamics and individuals' emotions.

In particular, here are the "blind spots" for the high hands styles (our adaptation):

- Executer
  - Expects others to follow through and becomes surprised and frustrated if they don't
  - Loses sight of the big picture and the overall mission at the strategic level
  - Goes into data overload or analysis paralysis, with a focus on too much detailed information not geared at the desired project outcome
- Facilitator
  - Fails to stay focused on the true project goals
  - Begins to feel isolated from everyone else on the team, without others properly knowing what you—in particular—contribute to the effort
  - May not involve the full set of external stakeholders
  - May neglect to value what they themselves are personally learning on the project, and may lose opportunities to parlay project success into "healthy self-promotion"
- Driver
  - Direct and straightforward communication style may come across as blunt or intimidating
  - Mode may prevent airing of opinions and alternatives from the rest of the team
  - Need to "keep score" can alienate the other styles who don't value "scoring" and "winning" as much as the driver
  - Pace of change and high expectations of driver can exhaust other team members causing decreases in the project team's quality and/or quantity of output
  - Can fail to include the team's dynamics on their radar screen
  - Can undervalue the needs and concerns of team members

So overall, the hands-based styles are very focused on building, own-ing, and managing the plan. It lines right up with PMI's balance of processes—more than half of the 47 processes in the PMBOK Guide (24 to be precise) are planning processes. The signature of this style is almost literally written into the "Standard" for the discipline. Hands people are also all about acquiring, building, and making the most effective use of resources—often borrowed from disparate organiza-tions in their enterprise, to get things done. Sound familiar?

This concept of "project managers as *Hands* people" significantly resonated for us when we looked at the characteristics of the *Hands* people. They are as follows: *Efficient, Tactical, Process Oriented.* Sound familiar? That's a pretty succinct way to describe PMs, and at the same time it illustrates a *problem*, and it is one of the main reasons we wrote this book. We realized that the Sustainability Wheel itself was—by definition—change. In our analogy, if the Sustainability Wheel represents "the rubber hitting the road," well, then by definition we're talking about the potential to move, or change.

Now back to the heart, head, and hands model. We've asserted that PMs are "hands"—"get things done"—people. By being focused on "getting things done" and getting them done well and on budget and on schedule, we tend to lose sight of the big picture and forget sometimes about team dynamics and the emotions of individuals, as summarized earlier. Eyes on the prize.

However, we probably never have our eyes on that prize as it yields continuing benefits in 2, 5, 10, and 100 years, and not on the way the prize fits into the whole *set* of prizes.

As Trautlein says, referring to project management certification programs, "these … do not deal adequately with Head … they often encourage a focus on the goals of a project, but the vision … is often missed. Limited attention is given to the overall business strategy, and the pivotal task of helping others see the connection between their project and the big picture is downplayed."

The book inspired us to create a graphic, based on the fact that PMs are "hands" people to illustrate the ways in which sustainabil-ity thinking feathers into project management as well as the multiple places where roadblocks exists. See later.

As you can see in Figure 1.11, a project begins with a Charter, hope-fully connected well to the mission, vision, and values of an enterprise.

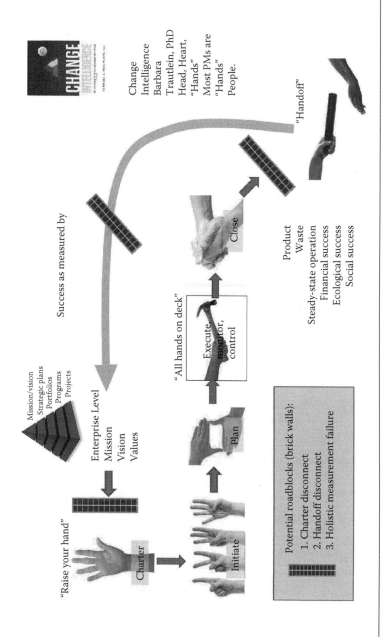

Figure 1.11 The hands have it.

That brings us to our immediate and critical first roadblock. That charter might NOT be connected well. In fact, we have many examples of cases where projects start immediately off on the "wrong foot" because of this disconnect. For example, consider a company with a very public image of ecological responsibility launched a product which produces a by-product that produces a nonrecyclable plastic piece that consumers must dispose of and are ending up in landfills in the billions. Economically successful, the product still caused the company significant embarrassment and a blemish on the company's name. In any case, the charter disconnect must be avoided, and can be if the recommendations in this book as well as those referenced in the previous research are followed.

The chartered project now moves through the process groups of Initiation, Planning, Executing, Monitoring and Controlling, and Closing. We use "hand" metaphors for all of these. At Closing, the project is "handed off" to operations. Here, we find our next roadblock. When is this handoff—when in time? What criteria are used? When does the PM really let go? And perhaps even more importantly, even though the PM is releasing the project's product to the steady state, has he or she let their thinking float forward to the time when the product is "doing its thing" in the long term?

Next, the product, in operation, is measured for success. Are all three aspects of the TBL being used in measuring its success? Does the enterprise look at environmental and social impact as the product is used? Are they willing to launch improvement projects to tune the product, if it is not? This represents the third roadblock. Here, we have to assure that the operations team is connected to the mission, vision, and values and is considering (and measured by) not only the economic bottom line but also the social and ecological elements of the TBL. Again quoting

Managing Change in Organizations: A Practice Guide, The ultimate purpose of change (i.e., a project) is to contribute to the organization's continued growth and to sustain its competitive advantage. Successful execution of the change can only be measured through benefits realization, which is an assessment of the successful integration of the change into business as usual ... where each expected benefit is aligned with the vision and its contribution to the ... purpose at the organizational level.

Change Intelligence at Various Hierarchical Levels in the Enterprise

The following is an adaptation of an article recently written by Dr. Barbara Trautlein and used with permission. In it, we can see the PM's adaptation of change in context with executives and with supervisors.

Leading Change Across Levels:

Tips to Apply and Traps to Avoid for Executives, Project Managers, and Supervisors

by Barbara A. Trautlein, Ph.D.

Change challenges vary by organizational level and role. For executives, the challenge is to lead the entire enterprise, transforming its traditional operating systems and organizational culture to be more competitive for the future. For the project manager (PM), the challenge is to design and implement a new technology with limited budget, staff, and authority. For supervisors, the challenge is to motivate the team to adopt new ways of working, even though they may not have all the training or tools to do so themselves.

Not only do leaders face different types of problems based on the types of changes they are tasked with, their challenges are also impacted by where they sit in the organizational hierarchy. But no matter what position you currently fill, you will be able to lead change much more effectively when you understand how Change Intelligence, or CQ, works at different levels of your organization.

Even if they are "open" and participative, most organizations are still structured hierarchically. Change leaders can exist at any level of the hierarchy, but there are predictable differences in how people at the top, middle, and bottom relate to organizational change. Those at the top usually set the direction of the change and are most convinced of the need for it, but they tend to be isolated from many of the change's direct impacts. Employees at the bottom, though they are most removed from the rationale behind the change, are often most directly impacted by it; an alteration in their behavior is usually a significant part of the change initiative, and they can thus appear most resistant to it. That means that supervisors and managers typically find themselves stuck in the middle, squeezed between these two levels, sandwiched between the edicts of their bosses and pushback from their staff. And PMs have their own set of change leader challenges. Executives initiate change and supervisors implement them. Similar to executives, PMs can influence an initiative's overall direction, but typically they're not yet strategic leaders. Similar to supervisors, PMs are accountable for executing change, but they have to operate on a more tactical level as they plan and coordinate the change process. This process typically involves people from multiple departments, working together on a temporary project team, whose members often report to other managers and who have additional and potentially conflicting responsibilities.

WHAT IS CQ AND WHAT DO WE KNOW ABOUT LEADING CHANGE AT DIFFERENT LEVELS?

Change intelligence, or CQ, is the awareness of one's own change leadership style and the ability to adapt one's style to be optimally effective in leading change across a variety of situations. Regardless of level or job title, each change agent has a basic tendency to lead with his or her heart, head, hands, or some combination of the three. Leaders who lead mainly from the heart connect with people emotionally (I want it!). Those who lead from the head connect with people cognitively (I get it!). And those who lead from the hands connect with people behaviorally (I can do it!). The powerful combination of all three is what Change Intelligence, or CQ, is all about.

Research indicates that executives tend to lead with the "head," project managers with the "hands," and supervisors with the "heart" (Figure 1.12):

Said another way, almost half of executives surveyed lead change by focusing on vision, mission, and strategy (head strengths). Almost 40% of project managers emphasize planning, tactics, and execution (hands' strengths). And more than half of supervisors place a premium on connecting, communicating, and collaborating (heart strengths). Executives tend to have their radars tuned to purpose, project managers on process, and supervisors on people.

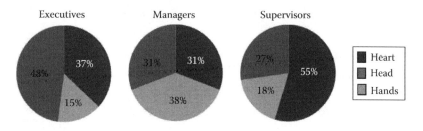

Figure 1.12   Executives, managers and supervisors lead differently.

Overall, these results are logical, and a good thing. A defining aspect of the executive role is to spearhead organizations toward brighter futures. Project managers are accountable for adhering to schedule, scope, and budget. And it's great news that admonitions for frontline leaders to engage in "coaching" and "motivating" their teams are embraced.

HOW CAN LEADERS AT ALL LEVELS USE CQ TO LEAD CHANGE MORE INTELLIGENTLY?

While all leaders have a natural tendency, and while certain roles may mandate a specific focus, the most effective change leaders at any level are able to flex their style when called for to manage successful and sustainable change.

Executive change leaders at the helm of an organization "engage the brain" to perform the critical function of scouting out new opportunities, discovering trends that could impact the business, and steering toward new horizons. However, at times they may neglect the map and the needs of the people whose help they need to realize their vision. Questions executive change leaders should ask themselves to avoid common head-oriented traps:

- While you imagine new possibilities, are you keeping your feet firmly planted in the here and now? How can you translate your lofty vision to specific plans and tactical steps so others can confidently champion them?
- What's the potential impact of your vision on the organizational culture? What do you need to do differently to ensure a committed and engaged workforce?

Project manager change leaders "help the hands" get things done, and their execution is usually backed up by comprehensive, step-by-step plans. While focusing on the details, such change leaders may neglect the big picture, and are prone to overlooking the need for positive team dynamics. Questions project manager change leaders should ask themselves to avoid common hands-oriented traps:

- Are you balancing execution with communicating the why of the change and where it's taking your team or organization? Do people focus on more than just today's to-do list?
- Do you make it a practice to set up structured time to meet with key stakeholders and ensure that they're on board with the direction your project is headed? Although your plan may be logical and sound, if you haven't addressed the concerns of key players, they may not be supportive when the time for implementation comes.

Supervisor change leaders "inspire the heart," engaging their team members and supporting the people around them as they all move through a change process. However, such change leaders may not confront others who are not behaving consistently with the change or give enough emphasis to completing tasks and making progress toward challenging new goals. Questions supervisor change leaders should ask themselves to avoid common heart-oriented traps:

- Do you shy away from giving constructive criticism because you think it might damage your relationship? In times of change, it is critical to both reward positive behaviors and give people constructive feedback when they were not performing to changing expectations.
- Do your team members challenge you because you skirted the rules and didn't adopt new work practices when you were their peer? Show vulnerability and build trust by admitting your mistakes, explaining why you were wrong, and committing to role model change-friendly behaviors in the future.

Change intelligent leaders demonstrate the savvy to apply all three tools in their tool bags—to engage the brain, help the hands, and inspire the heart—so people at all levels are empowered, equipped, and engaged to partner together toward mission-critical transformation.

More about Projects, Programs, Portfolios, Leadership, and Change

Helping us validate the connection between project management, sustainability, and change agency, Dr. Cynthia Scott, author of *Leadership for Sustainability and Change*,* uses a figure to show how different types of people in enterprises are open to change. In a scale that ranges from "No Commitment," where they may be passively or actively resisting change or denying reality of the need for change, to the opposite end, "Make It Happen," we can agree that the project managers, in general, are going to at least be in the 45% of the population that is open to change from a "Help It Happen" to "Make It Happen" perspective (see Figure 1.13).

For a person taking a lead in change management, the book provides a methodology for doing so, one that should look pretty familiar to us as PMs, in that it should evoke the Deming or Shewhart Cycle—Plan, Do, Check, Act. The model is called "Sense, Scout, Synthesize, and Steer." A model along with a description of each of the modes is given in the following.

- *Sense*: Identify your purpose. Connect your core values to your personal intention to become resilient and establish a foundation for leading change.
- *Scout*: Look around you. Understand and appreciate the stages of transition, assess organizational and individual readiness, and identify key people and leverage points for change. We would add: take advantage of whatever your enterprise key messages are about TBL thinking that they make available to the public and to other stakeholders. Sometimes this is as easy as going to the "About Us" tab or page of your enterprise's public website.

---

* C. Scott and T. Esteves (2013), *Leadership for Sustainability and Change*, DoSustainability, Oxford, U.K.

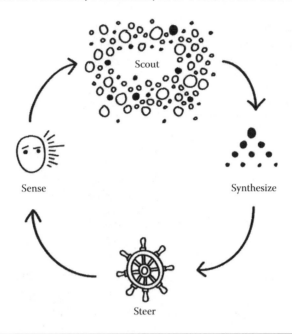

Figure 1.13 Sense, scout, synthesize, steer.

- *Synthesize*: Find patterns and build commitment. Map your change journey, design, test, and evolve approaches with rapid feedback, and engage others with stories of progress.
- *Steer*: Implement and calibrate. Mobilize action, track progress, encourage feedback, and continue to grow.

We have one more important idea to close this chapter—the idea of Purpose. Recall that the mission/vision/values—that gear at the top of our gear diagram which *drives* an organization—is *Purpose*. The element at the very top of the SEF is *Purpose*. Why is this important? According to Nikos Mourkogiannis, from his book, *Purpose: The Starting Point of Great Companies,** purpose* is crucial to a firm's success for three reasons:

- It's the primary source of achievement.
- It reveals the underlying dynamics of any human activity.
- It's what successful leaders want to talk about—they care about purpose because of what they see every day.

---

* N. Mourkogianniss (2006) *Purpose: The Starting Point of Great Companies*, Palgrave Macmillan, New York.

Further, Mourkogiannis says it all in one simple sentence: "Purpose is your moral DNA. It's what you believe without having to think. It's the answer you give when you're asked for the right as opposed to the factually correct—answer."

We'll revisit Purpose in some of the dimensions, particularly Detect, in upcoming chapters.

2
THE HUB
The Respect Dimension

Introduction

The "hub" of the Sustainability Wheel® represents the epicenter to which all of the other components are attached. The center—the hub—must be very strong in order for the wheel to smoothly support the bumps in the road and traverse the inevitable "road hazards" that could block our organization's path to a sustainable future. Respect is part of the hub. It is that connection between the corporate (organizational level) mission/vision and the organization's sustainability mission/vision.

Think of the Earth as a fairly delicate system—an egg comes to mind. A crack in that egg and the contents are liable to spill out, robbing the egg its chance to hatch and grow. Although organizations are getting the general message of sustainability, we sense a crack in this eggshell—a hairline crack, perhaps, but a crack nevertheless.

We have looked at literally hundreds of organizational mission statements. The crack we see is a "separation" between the sustainability goals and the business goals, rather than a fully integrated, triple bottom line viewpoint.

The majority of organizational missions/visions contain specific references to sustainability and sustainability efforts either undertaken or planned. We've also looked deeper into organizations to see if (1) those undertakings are having any overall effect, independently verified, and (2) those planned undertakings are coming to fruition. Some of the questions to answer are as follows:

1. Are those efforts connected to the organizations' overall corporate social responsibility (CSR), the overall organizational mission/vision?

2. Is there an environmental management plan (EMP)?
3. Is the EMP widely shared?
4. Is the EMP widely known throughout the organization?

When we are able to answer those questions in the affirmative, it gives us the hope that there are serious sustainability efforts preventing, or at least mitigating, the crack from going deeper into the shell. While there may be certain aspects of the damage we have to live with for a long time, it is some comfort that there are efforts to stem the tide, excuse the pun.

In this chapter, we will explore organizational commitment to sustainability by reviewing some ideal and less than ideal mission/vision statements and answering the four questions listed earlier. In subsequent chapters, we will connect those organizations with their success or their failure to implement sustainability efforts, including whether or not those efforts are embraced internally and externally, as well as how that perception affects the overall view of the organization from its stakeholder perspective.

The Mission Statement

What makes a good mission statement and how does sustainability connect with the overall mission/vision? From our standpoint (remembering the eggshell crack), we have often seen that there are *two* mission/vision statements for organizations. The first is the organization's overall mission/vision and the second is the vision for the organization's sustainability efforts. And yes, we assert that they should be fully integrated, but we suggest a slightly counterintuitive way to do this. We think it makes sense to first disconnect the two as separate efforts, but then later, after understanding each in their own context, then connect the sustainability efforts to the overall vision as a subsidiary plan (to borrow a phrase from project management) to the overall mission/vision. The reason for disconnecting and reconnecting is so that the sustainability efforts are not lost in the overall mission, can be rationalized if needed, and to emphasize the sustainability efforts as a major component to the overall mission, rather than an afterthought. Again, to borrow from project management, as well as our book *Green Project Management*,* "... think of the environment (sustainability), in the same way we think of quality. It must be planned in." By having a sustainability mission/vision and connecting it with the overall mission/vision, it becomes part of the plan.

A good organizational mission/vision includes the following.

According to Forbes online,[†] there are four questions that need to be answered to "Get a Good Mission Statement"

1. What do we do?
2. How do we do it?
3. Whom—or what—do we do it for?
4. What value do we bring?

These are good steps to follow for a *sustainability* mission/vision statement, too. Some of the issues we have found with some mission statements are that they are vague, incomplete, unfocused, and lack in the value statement, which we consider crucial, particularly in a sustainability mission/vision. Without stating the value of sustainability efforts, it is difficult to convince stakeholders to invest in the effort. A good place to start is to review your own organization's vision statement with those four questions in mind. It would also be good to review the vision statements of any companies in your supply chain to determine if *their* values are *your* values.

The following is a list of 10 things to look for in a good sustainability mission/vision statement. It includes revisiting and updating the statement to assure that it is relevant given current marketing (profit) conditions, sustainability "climate," environmental (planet) changes, employee and customer (people) demographics, and other changes that can potentially impact the "3 p's" and the organization.

Most importantly, the sustainability mission/vision statement must be easy to read and easily located on the company's website and its annual report. It also needs to be connected to the organization's

---

* R. Maltzman and D. Shirley (2011) *Green Project Management*, CRC Press, New York.
[†] http://www.forbes.com/sites/patrickhull/2013/01/10/answer-4-questions-to-get-a-great-mission-statement/ (accessed July 18, 2014).

overall mission/vision as well as to the organization's important sustainability documents for legitimacy and to show that there is buy-in. It needs to be tested with the organization's stakeholders. One way to assume that is to read the statement and see if it relates to you as a customer, employee, or other stakeholder.

Ten Things to Look For in a Good Sustainability Mission/Vision Statement. Is it:

1. Short and simple?
2. Specific to the company?
3. Visible and easy to find?
4. Connected to the overall mission/vision of the particular organization?
5. Sharing links to relevant corporate documents for transparency (i.e., your EMP)?
6. Tested with employees?
7. Tested with suppliers and partners?
8. Tested with customers and other stakeholders?
9. Evaluated as necessary to confirm relevancy?
10. Updated as necessary to keep it relevant?

The following are several case studies to illustrate what a good mission statement should look like and the reasoning behind these judgments. For us, the best way to analyze a mission/vision statement is to see how it relates it to the three "p's" of sustainability, *people, planet, profits.*

Patagonia—Case Study   In our first book, we acknowledged Patagonia as one of those companies that are "at the top of their game" when it comes to sustainability. Patagonia, and its leader, Yvon Chouinard, have been in the forefront of sustainability since the company was founded over 40 years ago. It has been widely reported that the company was founded because Mr. Chouinard wanted to protect the fragile mountain environment from the damaging pitons being used by climbers. He developed a piton that was less damaging. After demand outgrew his ability to produce his climbing gear by hand, Chouinard, in partnership with Tom Frost, began redesigning, improving and manufacturing climbing gear. For additional reading, Chouinard's book* discusses in detail, Patagonia's sustainability journey.

"Build the best product, cause no unnecessary harm, use business to inspire and implement solutions to the environmental crisis," is Patagonia's mission statement. While Patagonia does not have a specific mission/vision statement for their sustainability efforts, their overall mission statement *is* their sustainability mission statement. This statement certainly addresses the profit (best product) and planet (cause no unnecessary harm, use business to inspire and implement solutions to the environmental crisis). Patagonia specifically addresses the people the aspect of sustainability through their CSR. According to their website,[†] their mission is to "promote fair labor practices and safe working conditions throughout Patagonia's supply chain." To accomplish that, they take a three-pronged approach:

1. Working with their factories to promote fair labor practices and ensure good working conditions
2. Working with their mills to produce high-quality materials while reducing environmental impacts
3. Using their California SB 657 disclosure statement to outline the steps they are taking to monitor and assist their suppliers to meet human rights standards particularly in the areas of human trafficking and child labor

In summary, to give the overall impression of Patagonia's commitment to the people and the planet, it is worthwhile to provide an abridged version of one of the frequently asked questions; "How does Patagonia weigh its commitment to environmental versus social responsibility?" In their answer, Patagonia has given over $40 million in cash contributions to environmental causes, helped launch 1% for the planet, worked with factories and mills (as mentioned earlier) to reduce harm, and worked to improve lives and protect health as a founding member of the Fair Labor Association (FLA). Additionally, "We also take some unusual business actions to advance social responsibility throughout the supply chain. Our Social and Environmental

---

* Y. Chouinard and V. Stanley (2013) *The Responsible Company*, Patagonia, Patagonia Books, October 6, 2013. Accessed September 12, 2014.
† http://www.patagonia.com/us/patagonia.go?assetid=67372.

Responsibility (SER) team works in the Production Department, and with the Quality staff, not in the administrative or marketing arms of the company. All three teams (SER, Quality, Production) work and travel together. Each team's director has an equal say in sourcing decisions for new and current goods. Each has veto power over doing business with a new factory." While they do not publish a CSR report, their transparency can be evaluated through a social audit report from the FLA.[*]

Stonyfield Farms—Case Study   We began reporting on Stonyfield Farm's sustainability effort in our last book, *Green Project Management*.[†] Stonyfield's mission statement[‡] is a little longer than most. We believe that is because their sustainability mission and the corporate mission are inextricably combined. It is as follows:

- To provide the highest quality, best tasting, all natural, and certified organic products
- To educate consumers and producers about the value of protecting the environment and supporting family farmers and sustainable farming methods
- To serve as a model that environmentally and socially responsible businesses can also be profitable
- To provide a healthful, productive, and enjoyable workplace for all employees, with opportunities to gain new skills and advance personal goals
- To recognize our obligations to stockholders and lenders by providing an excellent return on investment

Stonyfield Farms do not separate their corporate mission from their sustainability mission. Reading through their mission, it is easy to see the connection. The first bullet is the overarching corporate mission: "To provide the highest quality, best tasting, all natural, and certified organic products." The next four bullets above address the people, planet, and profit aspects of their sustainability message.

---

[*] http://www.fairlabor.org/affiliate/patagonia (accessed September 25, 2014).
[†] GREEN PM placeholder.
[‡] G. Hirshberg (2008) *Stirring It Up*, Hyperion, New York, pp. 23–24.

From an article by Stonyfield Amy, *At Stonyfield, the Healthy Mission Came First, Yogurt Making Second*—(see more at: http://www.stonyfield.com/blog/about-stonyfield/#sthash [accessed September 25, 2014]), following their mission, the people at Stonyfield have "pioneered planet-friendly business practices—from offsetting our yogurt works' emissions, to making yogurt cups from plants instead of petroleum, to making our own renewable energy, and much more." And, they have been named one of the best companies to work for in 2010 and 2011, and received the 2010 Business of the Decade Award. In order to accomplish what they have to date, their "hub" is very strong.

General Motors—Case Study　According to their website (http://www.gm.com), their mission statement is as follows:

> G.M. is a multinational corporation engaged in socially responsible operations, worldwide. It is dedicated to provide products and services of such quality that our customers will receive superior value while our employees and business partners will share in our success and our stock-holders will receive a sustained superior return on their investment.

Connecting to that statement is their sustainability mission statement:

> We're committed to continuous improvement as we reduce the environmental impact of our vehicles and facilities. We're making progress—through vehicles like the Chevrolet Volt, our 111 landfill-free facilities and by receiving back-to-back EPA ENERGY STAR® Partner of the Year—Sustained Excellence awards.

Their "socially responsible operations" translate to reducing the environmental impact of their vehicles and facilities.

This is their "hub" for sustainability and is strengthened by further definition of their commitment to sustainability. Specifically,

a. "At GM we view sustainability as a business approach that creates long-term stakeholder value. It is an approach that is executed by every function at every level of our company."

b. "Sustainability is a value proposition that takes into consideration environmental, social and economic opportunities and supports the long-term success of the company."

c. Our sustainability strategy aims to create long-term stakeholder value; align corporate policies, positions, and sustainability initiatives; focus efforts on areas of significant impact; and be executed within every function by every employee. The strategic pillars are focused on four specific areas as follows:

1. *Innovations* that grow our business through new products and services that customers desire while addressing environmental issues and social concerns

2. *Integration* that ensures sustainability is embraced throughout GM

3. *Transparency* that builds trust and accountability

4. *Employee engagement* that encourages a sustainable mind-set at GM

d. Our dedication reaches further than compliance with the law to encompass the integration of sound environmental practices into our business decisions. Guided by our environmental principles, we consider the environment throughout all aspects of our business, from our supply chain, to manufacturing, to the vehicles we put on the road. These are the principles that help frame our planning and decision making for our company's future:

1. We are committed to restoring and preserving the environment.

2. We are committed to reducing waste and pollutants, conserving resources, and recycling materials at every stage of the product life cycle.

3. We will actively participate in educating the public about environmental conservation.

4. We will vigorously pursue the development and implementation of technologies to minimize pollutant emissions.

5. We will work with all government entities for the development of technically sound and financially responsible environmental laws and regulations.

6. We will continually assess the impact of our facilities and products on the environment and the communities where we live and operate with a goal of continuous improvement.

GM's sustainability mission/vision and accompanying documentation are a model for other organizations to follow. The mission/vision is clear and concise, while the explanatory information clearly identifies the goals and objectives for their sustainability program. They obviously have spent considerable time and effort developing their statement and strategy. Unfortunately, developing a "sustainable" sustainability program is neither quick nor easy. It takes a great effort, especially when you are the size of GM. We've always said about their planning effort for project management and it applies here; the size of the effort is commensurate with the size of the project (organization in this case). A smaller organization must put the effort in, but certainly not the same effort as a GM for example. As you can see, however, GM's "hub" is surely strengthened by this effort, which keeps the "Wheel" strong and on solid ground.

EarthPM

EarthPM is a small (two person) company that specializes in project management and sustainability training and consulting services. To be transparent, the principles (two people) of the company, Richard Maltzman and David Shirley, are the authors of this book as well as the authors of *Green Project Management*, CRC Press, 2010, Cleland Award winner for excellence in project management literature, 2011. We've included our own sustainability mission/vision statement to show that even a small company like ours should and did spend a considerable effort thinking about and documenting our sustainability mission/vision.

From EarthPM's website (http://earthpm.com [accessed September 29, 2014]), "According to PMI—the Project Management Institute, the world will spend 1/5 of its GDP on projects, a hefty U.S. $12 Trillion this year alone. *That's a lot of energy put into projects.* Those projects will use energy, save energy, use resources, and save resources in ways we cannot yet imagine."

This site is devoted to the intersection of project management and "Green"—where green has to do with preventing climate change, preserving resources, and getting things done effectively and efficiently, which should already be flowing in the "green" blood of any project manager worth their weight in risk registers.

EarthPM's Mission and Objectives

Mission Provide the critical link between project management and environmentalism to *increase awareness amongst project managers of the power they have to improve the greenality* and effectiveness of their projects* whether or not they are *directly* involved with the environment.

Objectives

- Seamlessly blend the discipline of project management with environmental aspects of projects (we assert that *every project has environmental aspects*)
- Use varying media to reach our stakeholders: project managers, environmentalists, business leaders, and, in fact, all residents of this planet
- Develop greenality processes to enhance the project managerial blend of project tools
- Use greenality to save resources, time, and costs of the project and those of the earth

Additionally, to strengthen our "hub," we include our five assertions of *Green Project Management* (Figure 2.1).

We focus on sustainability; therefore, our mission/vision *is* our sustainability mission/vision. One thing that is missing from our mission statement is how we, as a company, intend to run our operation as

1. A project run with green intent is the *right thing to do,* but it will also help the project team *do things right.*
2. Project managers must first understand the green aspects of their projects, knowing that this will *better equip them to identify, manage, and respond to project risks.*
3. An environmental strategy for a project provides added opportunity for success of both the project and the *product of the project.*
4. Project managers must view their projects through an *environmental lens.* This increase the Project Manager's (and the project team's) long-term thinking and avails the project of the rising "green wave" of environmentalism.
5. Project Managers must think of the environment *in the same way that they think of quality.* It must be planned in, and the cost of "greenality," like the cost of quality, is more than offset by the savings and opportunities.

Figure 2.1 The Five Assertions of EarthPM™.

---

* Greenality—The degree to which an organization has considered environmental (green) factors that affect its projects during the *entire project life cycle and beyond.*

sustainable as possible. Of course, because sustainability is our business, we operate with similar principles as GM, but for us, we might write our operational sustainability objectives (borrowing from GM) as follows:

1. We are committed to restoring and preserving the environment.
2. We are committed to reducing waste and pollutants, conserving resources, and recycling materials.
3. We will actively participate in educating the public about environmental conservation by continuing to include these issues in our teaching.
4. We will continue to purchase products for our business and personal use that minimize pollutant emissions.
5. We will continually assess the impact of our work and operate with a goal of continuous improvement for sustainability.

To reiterate, the "hub" is the epicenter that the rest of the Sustainability Wheel relies on for strength. A substantial effort is required to make the hub as strong as possible to keep the wheel rolling along and able to withstand the "road hazards" that will be encountered. Some of those hazards will be in the form of questions as follows:

- Why do we have to consider long-term effects, when our projects are temporary, unique, and have definite start and end dates?
- I receive my orders and do my job, why should I care about how the company chooses their projects?
- How will I know that my project is successful with regard to sustainability?
- How does sustainability affect the overall goals of my project?

Answering these questions is part of the reason that the mission/vision statement should be carefully constructed, clear and concise, and visible to all. A strong "hub" means a strong Wheel.

3

THE SPOKES

The Reflect Dimension

Introduction

The "spokes" of the Sustainability Wheel represent the channels that connect the mission/vision (strategic, sustainability, project) values to the "hub" of the Wheel. While the hub must be very strong in order for the Wheel to, quoting Chapter 2, "smoothly support the bumps in the road and traverse the inevitable 'road hazards' that could block our organization's path to a sustainable future," the spokes need to be strong and *tight* to ensure that the organization's mission/vision (Respect) and its mission/vision are "heard" (Connect). The reflect dimension tests to see whether organizational artifacts, including the organization's mission/vision, corporate social responsibility, sustainability mission/vision, sustainability roadmap, and environmental management plan (EMP), are being communicated, accepted, and practiced. It specifically targets those groups of people who *need* to be connected: project managers, program managers, and portfolio managers for instance.

Georg Kell, executive director, United Nations Global Compact, is quoted in a report from Accenture.com entitled "The Sustainable Organization: Lessons from Leaders Series—the Chief Executive Officer's Perspective." Copyright © 2012 Accenture. Mr. Kell says, "The material aspects of sustainability give a big competitive advantage to corporations that have long practiced and built up competencies at the strategic and operational levels." While this may seem to be more of a "Connect," it all starts within the organization. To reinforce this sentiment is the following quote from http://coso.org/documents/ (accessed October 3, 2014), "Integrating the triple bottom line into an enterprise risk management program," Ernst & Young LLP, Craig Faris,

Brian Gilbert, Brendan LeBlanc and Miami Brian Ballou, Dan L. Heitger, 2013; "For many organizations, sustainability has evolved from a 'feel good' exercise to a strategic imperative that focuses on economic, environmental, and social risks and opportunities that, left unattended, can potentially threaten the long-term success of strategies and the viability of business models. They understand that sustainability is not one function's domain, but rather a responsibility that the *entire enterprise* (author's emphasis) needs to own."

Environmental Management Plan

The EMP is a framework that can help direct an organization's efforts to achieve environmental goals and objectives. This is the "planet" part of the three p's, people, planet, profits; and may also contain elements of the other two P's, people and profits. Through this framework, an organization is better able to define, institute, and manage their long-term commitment to achieve those goals and objectives. Included in a good EMP are as follows:

1. A thorough review of the organization's environmental goals and objectives, tested against an objective standard.*
2. Analysis of the organization's environmental impact.
3. Review of any legal obligations as they relate to the environment, including adherence to the Clean Water Act of 1972, the Clean Air Act of 1970 and all the amendments, as well as standards and best practices like the Energy Star Program.
4. Set goals and objectives based on the analysis and reviews.
5. Use SMART† objectives so that they can be measured and monitored.
6. Connect with employees and other stakeholders.
7. Commit to continuous improvement of goals and objectives.

The federal government uses the term "environmental management system" (EMS) as opposed to EMP, but they are essentially the

---

* An example of objective standard could be a "best practice" in the organization's industry.
† *S*pecific, *M*easureable, *A*greed Upon, *R*ealistic, *T*imely.

Table 3.1 Costs and Benefits of an EMS

POTENTIAL COSTS	POTENTIAL BENEFITS
Internal • Staff (manager) time • Other employee time (*Note*: Internal labor costs represent the bulk of the EMS resources expended by most organizations) *External* • Potential consulting assistance • Outside training of personnel	• Improved environmental performance • Enhanced compliance • Pollution prevention • Resource conservation • New customers/markets • Increased efficiency/reduced costs • Enhanced employee morale • Enhanced image with public, regulators, lenders, investors • Employee awareness of environmental issues and responsibilities

same thing. Table 3.1 illustrates some of the costs vs. benefits of an EMS (EMP).

EMS and ISO 14001

Part of the focus of an EMP should be to encourage an organization's continuous improvement of its commitment and policies. Figure 3.1 illustrates the "cycle of improvement." There is a measurement part here. To be effective in their efforts, an organization *must not fail* to recognize that in order to truly succeed, the efforts need to be

Figure 3.1 The continuous improvement cycle.*

* U.S. EPA, http://epa.gov/ems/, (accessed October 3, 2014).

monitored (evaluated) and reviewed. That part of the process is just as important as the commitment itself. Without monitoring and reviewing the policy and commitment, there is no way of knowing whether or not they are effective. What is the use of making a policy and a commitment and not following through? Stakeholders can easily see whose commitment is real and whose commitment is just "talking the talk." Stakeholders will have a tendency to trust the organization that makes a real commitment to an environmental policy and follow up with evaluation, review, and change, if necessary. Realistically, change will be necessary because the conditions surrounding the organization are always in flux. Just on climate change alone, new research is appearing almost daily.

Figure 3.1 illustrates the framework of an EMP developed by the International Organization for Standardization (ISO), specifically ISO 14001 standard. Obviously, the cornerstone of any system is the commitment of an organization's management. This commitment is the establishment and guarantee that the policy developed is recognized as the standard for the organization. Acknowledgement of this commitment needs to be a strong message from management. While there are different ways to organize, top-down seems to always be the way important organizational messages are communicated.

The purpose of planning is to identify any environmental aspects of its operations, particularly those aspects that may have negative environmental impacts on the planet and people. The third "P," profits, will play a part in the overall ability of an organization to implement its policies. Those impacts could be positive, in that efficiencies gained can be translated to the bottom line, or negatively in either fines, loss of credibility, or actual expenses to implement the policy.

Some of the questions to be asked during the planning phase are as follows:

1. Are we asking the right people to determine the environmental aspects of concern?
2. What are the environmental aspects concerning our organization? (i.e., hazardous waste, water pollution, and caustic chemicals used in processes)
3. Have we clearly defined each concern?

4. What is the significance of each concern?
5. What should be our target of improvement for each of the highest ranking concerns?
6. What do we need to do to minimize those concerns?
7. What is the financial impact to minimize those concerns?
8. Can we measure improvement?

Once those questions are answered, a plan to address the major concerns is developed. The plan then should go through a review process to make sure it addresses the concerns' outline in the planning phase. Once that review takes place and is "blessed" by management, the next step is implementing the plan. Implementing will require committing the necessary resources to *effectively* execute on the plan. It will also include a way to measure and capture the results of the implementation. "Lessons learned" or some other documentation should be available so that all participants can record their observations. Those observations will then be reviewed along with any other developed effective measures and will be crucial in the "continuous improvement" process. Remember, there is nothing worse than a stagnant policy.

Enterprise Level

To be honest, the EMS is the mission/vision/values all rolled up into a *commitment* to do something. The organization's mission/vision/ values and its sustainability mission/vision/value are meaningless without a commitment to those mission/vision/values. Figure 3.2 illustrates the "flow" of the different levels. The enterprise level of an organization can be defined as an organization's capability and competency to execute on its commitments. *The Guide to the Project Management Body of Knowledge* (*PMBOK® Guide*), Fifth Edition, Project Management Institute, 2013, provides some insight into the capability and competency of an organization by referring to "Organizational Process Assets" and "Enterprise Environmental Factors." According to the PMBOK, organizational assets include "processes and procedures." Those processes and procedures are further broken down into (1) "initiating and planning," the "guidelines and criteria for tailoring the organization's set of standards processes and procedures, specific organizational standards such as policies, and

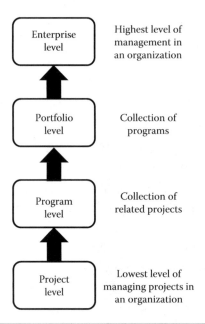

Figure 3.2 Hierarchical flow.

templates" (p. 27), (2) "executing, monitoring and controlling, change control procedures financial controls, issue and defect control, organizational communications requirements, risk control procedures," as examples (pp. 27–28), and (3) "closing, project closing guidelines or requirements" (p. 28), including lessons learned. Organizational process assets also include "corporate knowledge base; configuration management, financial databases, historical information" (p. 28); and other informational databases.

Enterprise Environmental Factors (PMBOK® Guide, Fifth Edition, pg. 29) include, but are not limited to:

- Organizational culture, structure, and governance
- Geographic distribution of facilities and resources
- Government of industry standards (e.g., regulatory agency regulations, codes of conduct, product standards, quality standards, and working standards)
- Infrastructure (e.g., existing facilities and capital equipment)
- Existing human resources (e.g., skills, disciplines, and knowledge, such as design, development, legal, contracting, and purchasing)

- Personnel administration (e.g., staffing and retention guide-lines, employee performance reviews and training records, reward and overtime policy, and time tracking)
- Company work authorization systems
- Marketplace conditions
- Stakeholder risk tolerances
- Political climate
- Organization's established communications channels
- Commercial databases (e.g., standardized costs estimating data, industry risk study information, and risk databases)
- Project management information system (e.g., an automated tool, such as scheduling software tool, a configuration management system, an information collection and distribution system or web interfaces to other online automated systems)

While factors can be considered in the context of connecting with an organization's sustainability efforts, so of the more important questions to ask at the stage are as follows:

1. Does our organizational culture, structure, and governance support our sustainability efforts?
2. Are we meeting all regulatory agency regulations?
3. Are our working conditions in compliance with all local, state, and country regulations?
4. Are our working conditions in compliance with our own codes of conduct, quality, and working standards?
5. Do we have the necessary personnel skills to execute and administer our sustainability efforts?
6. Do we consider the sustainability political climate in the areas in which we do business?
7. Do we have communications channels in place to effectively communicate our sustainability efforts to our customers?
8. Do we have communications channels in place to effectively communicate our sustainability efforts to our employees?
9. Are we using those channels effectively?

Portfolio Level

In the scheme of things, the organization's portfolio of projects is the collection of projects that may or may not be related to each other, but that forward the organization's mission/vision. In our previous book, *Green Project Management*, we said that nothing gets done in an organization without projects and that projects are where the "rubber meets the road" and "where ideas become real." Think of the portfolio level as the "filing cabinet" of all of the organization's projects. It is one step removed from the enterprise level. A critical piece for the enterprise is how it manages the portfolio or project portfolio management (PPM).

While the portfolio of projects is ideally structured in a way to achieve business results (mission/vision), it must be managed in a way to maximize the benefits of organizational assets, and more importantly for our discussion, the EMS (and overall sustainability). As part of that management, difficult decisions have to be made with respect to strategic prioritization. And, sustainability must be considered in that decision as an important part of the overall strategy.

Some of the questions that should be asked at the portfolio level are as follows:

1. Does the portfolio of project meet the strategic needs of the organization with respect to the mission/vision, especially the sustainability mission/vision?
2. Is the portfolio of projects being managed in a coordinated way to achieve the mission/vision?
3. Is the sustainability of the portfolio of projects being considered in the overall organizational decision-making process?

Program Level

The program level is the collection of related projects managed in a way that leverages the relationship between projects to maximize the efficient usage of resources. Those resources include environmental (planet), business (profits), and human (people) resources. Combining those three for our purposes is the sustainability relationship. Once again, it is important to view the different programs, and there may be several within a portfolio, in a way to maximize the benefits of an enterprise's organizational assets with respect to the overall sustainability as well as to help the enterprise make more strategic decisions. Questions that should be asked at this level are very similar to those asked at the portfolio level, except the focus is narrower. Specifically, the focus is more on those related projects without consideration of the effect on the portfolio.

1. Does the program meet the strategic needs of the organization with respect to the mission/vision, especially the sustainability mission/vision?
2. Is the program being managed in a coordinated way to achieve the mission/vision?
3. Is the sustainability of the program being considered in the overall organizational decision-making process?

While these questions may seem redundant, they act as a double check of how the portfolio is being managed. If the answers to the portfolio level and the program level differ, then it may be an indication that one or the other level is not being managed with sustainability.

Project Level

This is "where the rubber meets the road." Project management is the lowest level of the hierarchy, but most important to the organization. While that may look like the bottom of the food chain, in fact, it is the foundation that the sustainability of an organization is built upon. Without a strong foundation, i.e., vital connection to the sustainability mission/vision of an organization, the ideal will not be achieved. We've always asserted that projects are where ideas become real. For those ideas to contain the elements of the sustainability message of an organization (the ideal), the elements have to be present, or at least considered, for each project. Without that *connection*, the organization's sustainability efforts can be looked at as more "lip service" or, perhaps, as "green washing." Let's look at it from a *disconnect* point of view.

By not connecting sustainability at the project level, the foremost disconnect will be with project risk. Claiming ignorance of the sustainability risks in a project that have not even been considered is putting the project in critical jeopardy. Sustainability risk consequences include, but are not limited to the following:

- Purchasing from unethical vendors/suppliers
- Lawsuits
- Fines
- Loss of credibility

Because it is such an important connection to make, we'll look deeper into the sustainability risk consequences. According to an article, *What is Sustainability Risk?* posted by ENVELOLOGIC (http://envecologic. com/2012/07/11/what-is-sustainability-risk/ [accessed October 16, 2014]) on July 2012, "The starkest example that illustrates this scenario (how sustainability risk applies to business) is rising sustainability risk across sectors of an economy to varying degrees because of climate change." We've always asserted the following and it is confirmed by the ENVELOLOGIC post that continues "Whether or not climate change is leading to significant disorders is not even a point of debate. Innumerable agencies, including the *plus importante* Intergovernmental Panel on Climate Change (IPCC) have well established the adverse impacts that range from rising temperature to increase occurrences of extremes condition of flood and drought. These climatic disturbances have already been directly impacting the profitability of corporations." It is a risk to your organization.

The post continues to discuss the "financial risk, political risk, investors risk" to organizations. When we look at political risks, two things come to mind: government regulations, external effects on the organization, and the politics of image or reputation, both an internal and an external effect. While the regulation aspect can affect the real bottom line in the way of penalties, the image can, and will, affect the *value* of the business. The ENVELOLOGIC post uses a great example: "The Government of India levied coal premium of about $1 per ton of coal produced, giving a blow to the profit margins of coal producers and raising prices for power companies."

Continuing from that same post,

Large corporations have already been engaging in risk assessment and management. For example, PepsiCo has invested ahead of the curve to manage sustainability risks linked to water scarcity. As a critical raw material, water impacts input costs, competitiveness, and the ability to maintain production as well as influencing community relations and brand image. In 2009 PepsiCo announced 15 global goals and commitments focused on the sustainable use of water, land, energy and packaging. The firm aims to reduce water usage intensity by 20% between 2006 and 2015 across all manufacturing operations. It's high time that even medium and small scale companies follow suit and start paying serious heed to the sustainability risk, assess it and make plans to manage the risk if they want to ensure that growth takes place at an accelerated rate.

In a 2013 report published by the Committee of Sponsoring Organizations of the Treadway Commission (COSO), (http://coso. org [accessed October 15, 2015]), *Integrating the triple bottom line into an enterprise risk management program*, by Ernst & Young LLP; Craig Faris, Brian Gilbert, Brendan LeBlanc and Miami University; Brian Ballou, Dan L. Heitger. The publication highlights sustainability's "evolving" role in business, looking at the sustainability lens (an expanded view from the green lens in our previous book (*Green Project Management*) and its integration of sustainability into mission/vision. In addition, the post provides tips for raising sustainability awareness in an organization as well as looking to the future. It is all part of making that connection.

At the project level of an organization's sustainability efforts, several questions can be asked that are as follows:

1. Have I been given the necessary responsibility to execute on project-level sustainability efforts?
2. Do I have the support of my management?
3. Do I consider all aspects of sustainability in both the product of the project and the processes of the project?

Sustainability Programs and Incentives

Practicing sustainability counts. Whether it is a personal goal or contributing to an organization's bottom line, practicing sustainability should be rewarded. It is understood that there is an altruistic intention for practicing sustainability and that may be the only encouragement and reward needed. The French Philosopher Voltaire is quoted as saying "The biggest reward for a thing well done is to have done it." It is also good to be "officially" recognized for efforts related to a "job well done."

A formal sustainability program, widely distributed, is the best way to communicate to an organization that upper management is serious about their sustainability efforts. All of the right words in mission/vision statements won't count for much unless those words are well understood by everyone in the organization. The best way to do that is through the organization's formal sustainability program.

A sustainability program should, at a minimum, contain the following elements:

1. A centralized method to collect sustainability data.*
2. A method for analyzing and storing the data.
3. The data should be in a usable form in order to export it to analysis tools.†
4. A method to access and display the analysis.‡
5. A mechanism to communicate that the information is available.
6. An action register and lessons learned documents to address any of the issues that may arise and to keep a historical record.

It will also include a purpose statement to help gain acceptance and support. It is important to initiate relationships with the internal stakeholders. To effectively develop the program, input should be solicited from staff. But even with the input and buy-in of the staff, item 5 listed earlier continues to be extremely important to the success of the program.

The Green Carpet Award is a great example of academia formally recognizing "the outstanding efforts of teams and individuals across Harvard to create a healthy, more sustainable campus," according to their website http://green.harvard.edu/campaign/harvard-green-carpet-awards (accessed October 23, 2014). The program accomplishes two tasks: (1) recognizes outstanding sustainability efforts of student, both individual and teams, and (2) provides a forum for students to get together to discuss the issues to make a more sustainable campus. Again, according to their website, there are seven criteria:

1. *Innovative/Creativity*—seeks out and utilizes new technology; develops and advances better processes to produce impactful change.
2. *Replicable Models*—a "first of its kind," materials and resources from this project are made available to others across in order

---

* Sustainability data include energy usage information, resource consumption, environment, and health and safety issue occurrences.
† The analysis tools may be as simple as an Excel spreadsheet.
‡ Simple dashboards can be created.

to create efficiencies, streamline processes, and grow sustainability projects on campus.

3. *Collaboration/Engagement*—actively engages stakeholders within and beyond their school; engages fully in Harvard's sustainability programming, including Green Office, Green Loan Fund, working groups, and events.

4. *Energy/Greenhouse Gas Emissions Reduction*—creative opportunities that reduce energy and/or greenhouse gas emissions on campus and in buildings through construction, operations, engagement, technology, or new processes.

5. *Waste Reduction*—exemplifies "reduce, reuse, recycle and compost" properties in an office, buildings, or construction settings.

6. *Water Reduction*—explores creative opportunities to reduce water consumption on campus and in buildings through operations, engagement, or technology.

7. *Above and Beyond*—not only embraces sustainable practices such as the Green Building Standards, LEED, Green Office, temperature policy, etc., but continually strives to achieve more than is designated in these practices.

The Harvard Office for Sustainability has awarded the Harvard project team for the Massachusetts Green High Performance Computing Center (MGHPCC) the 2014 Green Carpet Award. The award recognizes Sustainability Leaders at Harvard who exemplify innovation and creativity through the development and execution of campus sustainability plans that focus on the reduction of energy use, greenhouse gas emissions, waste, and water consumption.

The Harvard project team that managed the MGHPCC was a recipient of a Green Carpet Award. Working for Boston University (BU), one of the schools involved with the consortium of universities collaborating in the MGHPCC, I am very familiar with the project. Every year I invite the BU project manager to present some information to my students for my Green IT course. It is a fantastic project and the award was well deserved. It not only provided the team with recognition, but continues to provide internal motivation.

In 2011, the National Environmental Education Foundation (NEEF) conducted an interesting survey highlighting the results of internal organizational efforts for employee engagement. The study, 2009 Engaged Organization Study available at http://www.neefusa.org/BusinessEnv (accessed October 22, 2014), reviews several companies' successful efforts to engage employees to be more sustainable oriented.

- *Cisco*—uses a dedicated intranet site with video and discussion forums. There is also an annual worldwide awareness campaign. The company uses "mixed-media and multidepartmental leadership" to get the message out to employees.
- *Hewlett-Packard* (*HP*)—"provides employee environmental education through mixed-media communications, events and programs." One example is the "brown-bag" informational seminars that include subjects like solar roofs and provides incentives for employees wishing to install solar roof panels. It has a Corporate Sustainability Group and also supports employee-led "green teams."
- *Interface Global*—is a company that we featured in *Green Project Management*, CRC Press, 2011, as "Top of Their Game" for sustainability. They are on a mission to reduce their environmental footprint to zero by 2020. Sustainability is a part of everyday for employees. Interface provides incentives and awards for sustainability efforts and ideas, as well as provides an online training in sustainable practices to all employees.
- *Stonyfield*—is another company that we featured in our 2011 book. It has always been a leader in the sustainability effort, providing a "Mission Action Plan" engaging all employees with Stonyfield's long-term sustainability goals. Using multi-departmental leadership, employee "green teams," mixed-media communications, and incentives, Stonyfield successfully gets the message out to employees.
- *Johnson & Johnson*—2010 was the target to get employees more engaged in sustainability efforts. Within the first year, 92% of J&J's facilities had an "environmental literacy plan" and "79% of facilities had deployed an annual environmental literacy module." The company has a corporate environmental health office and supports their efforts using mixed-media communications and multi-departmental leadership.

- *Wal-Mart*—is another company featured in our book, although primarily for achieving a "green" supply chain. However, they do have a "Personal Sustainability Project (PSP)" engaging more than 500,000 employees in voluntary sustainability efforts. Additionally, Wal-Mart provides a Sustainable Value Network (SVN) for its salaried employees. It offers mixed-media communications and multi-departmental leadership and support to "employee-led" green teams, reinforcing its program through performance incentives.

The NEFF updated this survey in 2013 and found some new and interesting results. Again, according to the website,

- "Sustainability" remains the established phrase to describe a company's environmental sustainability initiatives. "Greening" for many years was the second-most-used term but is now almost the least used to describe these initiatives.
- Social and environmental activities converge. As companies begin to address more complex supply-chain issues, those surveyed see environmental and social issues becoming more connected.
- Has sustainability knowledge become less important or have we "arrived"? In large companies, those surveyed see less of an increase in the value placed on a job candidate's sustainability knowledge than in years past, while mid-sized and small companies still see this as increasing.

Chapter 2 highlighted some of the efforts of Interface and Stonyfield, and the following case studies (used with permission from the National Environmental Education Foundation, David Lanham, Communications Manager, National Environmental Education Foundation 4301 Connecticut Avenue, Suite 160, Washington, DC 20008), illustrated in depth, provide significant lessons learned, relative to the positive effects of *reflecting* sustainability values within an organization. Contained within the case studies are important lessons to help make your organization more sustainable.

"Ray"sing the Bar for Sustainability

Case Study: Interface, Inc.
Company: Interface, Inc.
Industry: Commercial and residential floor coverings
Headquarters: LaGrange, GA.
Number of Employees: 3701
Total Revenue: $1.1 billion
Interface's Story
Joyce LaValle, Senior Vice-president of Associate and Customer Engagement

Synopsis: Interface, Inc. (Interface) is on a mission to have zero environmental footprint by 2020. To realize the mission, it has made sustainability part of every employee's job. Interface uses several educational practices including job appropriate training, storytelling, and learning modules to support employee learning and development.

Why the Program Was Started: An Epiphany at the Top of the Company Sustainability learning began at Interface in 1994 when our Chairman and founder, Ray Anderson, had an epiphany—an awakening to the importance of environmental issues and their relevance to Interface. This epiphany at the top of the company required extensive learning throughout the company to transform the organization according to Ray Anderson's vision.

How the Program Works: It's Part of Every Employee's Job There is nobody at Interface with a sustainability title; it's part of every employee's job. To operationalize the new direction, we developed a global advisory group to study sustainability and create a holistic plan for the company. Ray wrote a book about the plan, *Mid-Course Correction*, which was given to every employee. In the book, Ray lays out the underlying "why" as well as the entire plan. It is a great tool to help people to think about reengineering a company to be more responsible. Since then, every year, there have been thousands of people at Interface working toward the plan through thousands of projects. To support this mission, we see several educational practices, including job appropriate training and storytelling, and we're just beginning to create learning modules. We began early on with empowerment training that included outdoor experiences. The goal was to open people up to expressing ideas and thoughts, to provide room for anybody to question anything. We also have provided employees "Natural Step" training. Beginning in January 2008, InterfaceFLOR, the modular carpet division of Interface, Inc., introduced a new program designed to better introduce new associates to the company's culture. Entitled "The InterfaceFLOR Associate Experience," the site contains learning video modules or segments including Welcome to Interface, Our Story (history), Our Customers, Our Mission, Our Process and Product, Our Community and Our Culture: Your Role at Interface.

Key Lessons

- Make E&S part of a shared vision and the company culture, not a "flavor of the month"
- Measures are critical and the best teacher
- Storytelling is a powerful tool
- Include all employees
- Consider E&S motivation and knowledge in the hiring process

In the modules, associates can see and hear directly from InterfaceFLOR's senior management about our processes and culture, as well as from some of our associates about what it's like to work at InterfaceFLOR. Associates also can hear from local community leaders about Interface's support in the local community, and they can hear from several of our clients about our customer service approach. The pages are designed to be interactive and informative and are being used as part of a larger revamped orientation process for new associates, as well as a great learning tool for all associates.

We also do a lot of face-to-face meetings with employees: monthly meetings with all manufacturing team members and quarterly team meetings with others. The meetings are intended to communicate progress, share technologies and breakthroughs, and talk about future plans. Embedded in all of the conversations is discussion of progress related to sustainability and innovation.

All manufacturing team members are on a bonus program related to sustainability. Once a year, we hold a ceremony recognizing several business units for progress along several fronts, including sustainability. The site that makes the most progress encounters a tougher year of improvement ahead because we expect continual improvement! We have also started to use a StrengthsFinder survey administered by The Gallup Organization (the survey allows associates to identify their top natural talents that can be developed into strength) for our hiring process; we screen prospective employees to see if there is a good fit. *We have more people, young people, applying to the jobs at Interface than imaginable—and they care about the mission! Sustainability is often the first conversation we have during the hiring process. People want to come to a job with a purpose.* We had developed and worked on our plan for 12 or 13 years before we realized that we had never really talked deeply to the outside world about our sustainability vision. So, as part of a global exercise, we branded all the pieces of our sustainability vision "Mission Zero®." It's not really a brand—it's a promise that by 2020 we will have no adverse environmental impact on the world. Stated another way, we aim to have zero footprint: every creative, manufacturing, and building decision we make will move us closer to our goal of eliminating any negative impact our companies may have on the environment by the year 2020.

This was our first time to step out and be really bold about the sustainability vision publicly. At this point, we know we are so invested that we will continue this until 2020. We wanted the world to know about our vision so that we are held accountable by the outside world as well as by ourselves. The Missionzero.org website is being developed to help others join our mission—everyone is part of this Missionzero. org is also an educational tool for our employees, a way to have a conversation about the promise and the clarity of the vision. In this and other ways, we are trying to keep the sustainability vision front and center everywhere at Interface. There is nobody at Interface with a sustainability title; it's part of every employee's job.

Tangible Results: Interface Employees Neutralize Personal Travel Emissions
One of our most successful sustainability programs for our associates in Troup County (the county of Interface's headquarters)—Cool CO2mmute™—was developed from an idea by one of our manufacturing associates—Lina Marshall. During a company meeting,

Lina asked if there was a way non-sales associates could offset their travel emissions and Cool CO2mmute was born. Through this program, Interface associates make a voluntary personal commitment to reduce environmental impacts associated with their commute. By making a one-time yearly donation that the company matches, Interface and its associates neutralize personal travel missions through the purchase and planting of trees through American Forests. For all Interface business units in the Americas more than $10,000 was donated to American Forests in 2008 to sponsor the planting of trees that will result in more than seven million pounds of CO_2 being absorbed over their lifetime.

Measuring Results: Material Usage and Employee Engagement We have developed goals and measurements associated with our overall plan. We measure all materials coming in and going out, energy usage, yarn usage, etc. Every measure has some relationship to creating a sustainable company and is communicated for use in departmental impact plans. *We find that measures are, in themselves, critical and the best teacher.* In addition, starting about 5 years ago, we began measuring employee engagement throughout the organization, using a survey based on the Gallup book *First, Break All the Rules.* It includes 12 questions based on positive psychology that Gallup has developed for companies that want to be exemplary. If a company receives a high rating on these questions, it reflects deep engagement with employees. It is not a satisfaction survey nor is it specific to sustainability or Mission Zero. Sometimes we have areas that we need to work on, but in general, the scores show that Interface associates are deeply engaged with our sustainability mission and vision.

Challenges: Addressing the "Big" Questions Of course, we also encounter challenges: how do we keep making progress on our vision given the economy? How do we get off the grid affordably? How do we find the money to put reclamation technology throughout the world to get us off virgin materials? These are just some of the big questions.

Our observation is that on the journey there are periods of struggle where it appears you can't make progress and then the breakthroughs come rapidly! *We can't achieve our goal of zero impact without employee engagement. All of the innovation comes from employees.*

Advice for Others: Culture Change is Key   Educating employees about sustainability is important. But education alone can't make a company sustainable. What is really needed is a cultural shift and enormous commitment throughout the company. *Sustainability can't be a "flavor of the month."* If employees learn about sustainability but the company doesn't have a plan and vision, they are likely to become cynical and then the education will be a waste of money for the company. *If you turn people on to sustainability there needs to be a plan because they will take it up!*

Links to More Information

Interface website on Mission Zero: www.interfaceflor.eu/internet/ web.nsf/webpages/528_EU.html.

Mission Zero Network: www.missionzero.org.

Ray Anderson on sustainability on You Tube: www.youtube. com/watch?v=RcRDUIbT4gw.

Gallup's *First, Break All the Rules*: gmj.gallup.*com/content/1144/ First-Break-All-Rules-Book-Center.aspx.*

Making Sustainability Personal

Case Study: Wal-Mart Stores, Inc.
Company profile:
    Company: Wal-Mart Stores, Inc.
    Industry: Retail
    Headquarters: Bentonville, AR
    Number of Employees: more than two million
    Net Sales: $401 billion
Wal-Mart's Story
    Janelle Kearsley, Director, Private Label Sourcing
    Miranda Anderson, Director of Corporate Affairs, Sustainability
    Candace Taylor, Director, Strategy & Sustainability

Synopsis: Wal-Mart, the world's largest retailer, has the size and scope to influence change throughout the industry. Wal-Mart believes that its associates are critical in the company's efforts to become a more sustainable business. Its PSP has engaged more than 500,000 associates voluntarily in its sustainability efforts, demonstrating measurable results in associates' lives and in the workplace. Additionally, Wal-Mart's SVNs engage salaried associates, and all salaried associates' performance evaluations include a question on the associates' contribution to sustainability.

Why the Program Was Started: By Wal-Mart Associates, for Wal-Mart Associates As a global retailer, we recognize that we have a unique opportunity to participate in positive and sustainable change throughout the supply chain and into the homes of millions of people. This gives us the ability to fundamentally shift the way products are sourced, manufactured, delivered and sold. We believe that we can be a good steward of the environment and a good neighbor to communities around the world while growing a profitable business. We also believe that we can do this while saving our customers money so they can live better.

Key Lessons
- Make E&S outreach personal and voluntary
- Grassroots and personal involvement is essential
- Engage employees in setting goals

As a company, we are working to be supplied 100% by renewable energy, produce zero waste to landfill from our stores, and sell products that sustain our resources and the environment.

Using an approach we call Sustainability 360, we are working to achieve these goals and bring sustainable solutions to our roughly 61,000 suppliers, 176 million weekly customers around the world, and more than 2 million associates. Sustainability 360 lives within every aspect of our business, in every country where we operate, within every salaried associate's job description and extends beyond our walls to our suppliers, products, and customers. Overall, Sustainability 360 is about doing better for our customers, our associates, our suppliers, and our environment—and doing it together. Within Sustainability 360, we are engaging our more than 2 million associates in our sustainability efforts through two parallel avenues that involve listening to and working closely with our associates. Our associates around the world have the ability to educate their friends, family, and communities on sustainable practices. We are educating them on the environmentally friendly and ethical products that are on our shelves so they can, in turn, teach our customers about those products. In 2007, we introduced a voluntary, grassroots program called the PSP to all of our U.S. associates. PSP guides associates as they integrate environmentally friendly and healthy practices into their lives and make choices that benefit their communities. The program was started by Wal-Mart associates for Wal-Mart associates. It enables our 1.4 million U.S. associates to live healthier, more sustainable lives at work, at home, and in their communities. The program has grown to include associates around the world and has been successfully implemented in Brazil, Canada, China, and Mexico. Additionally, for our salaried associates, we have engaged them in our SVNs and added an additional sustainability component to their annual reviews.

PSP: Program Design

(Sustainability) relates to every single associate at Wal-Mart ... If we have 2.2 million associates worldwide, I'd love for every associate in every country to really recognize their personal, individual responsibility in the area of sustainability.

–Mike Duke, CEO of Wal-Mart Stores, Inc.

When we began developing the PSP program, we wanted to make sure that we were building a program that would be meaningful to our associates. We started by asking associates, "What does sustainability mean to you?" Their response was clear; sustainability had to be personal and relevant in their everyday life. From this insight, we developed the three main principles behind the PSP program:

1. It has to be personal and relevant.
2. It has to be voluntary.
3. It has to be bottom-up.

There are five general focus areas for PSPs:

1. Sustainable purchasing
2. Waste reduction
3. Health and wellness
4. Energy
5. Clean air and water

The goal for our PSP program design is to bring small change into the lives of our associates that can improve their own well-being as well as the health of the environment and communities.

PSP goals are chosen using the "SMART" goal framework:

- "S"—sustains the planet
- "M"—makes them happy
- "A"—affects the community
- "R"—regular and continuous in daily life
- "T"—takes visible actions that can be shared with others

PSP Implementation   Working with a consulting firm, we were able to develop associate education and training materials, determine a rollout strategy, and begin implementing the program into stores across the United States. Our first step in the implementation process was to have each store select two volunteers to attend a day-long, paid training session. Initially, our human resources department paid for headquarters training and the pilot training and the operations team paid for operations staff training. These retreats were held in natural settings, such as a state park. In the morning, we would lead participating associates through a discussion that focused on sustainability and Wal-Mart's environmental goals. In the afternoon, the discussion turned to educating associates on the PSP, how to develop a PSP and how to encourage others to join the program. The first training session reinforced the importance of all PSPs being voluntary, personal in nature, and focused on the local community. At the end of the session, each participant developed their own PSP—a small change that will help benefit their own life, and the health of the environment and local community—such as biking to work, quitting smoking, or losing weight. After completing the training, each volunteer became a PSP captain and was given the challenge to recruit 10 co-workers and train them to introduce PSPs to other employees. Through this design, captains are able share stories and the definition of a PSP and then let associates decide for themselves whether to adopt a PSP. Currently, we have approximately 46,000 PSP captains (about 10 per store and club), and host 120 retreats throughout the year. Captains serve as the sustainability advocates for our company in each store or club. In addition, in-store TV clips profiling employees and their PSP stories and a PSP-focused magazine complement PSP captains' efforts.

PSP Results: A Truly Grassroots Sustainability Movement   The project was piloted in Denver, Indianapolis, and Tampa in 2006, and rolled out to our Bentonville headquarters and Sam's Club stores in 2007. It is now deployed to 4,000 stores, and more than 500,000 associates have developed PSPs. The program has about a 50% rate of acceptance. Currently, we track the number of associates who adopt a PSP, the type of PSP and associates' success in aggregate.

As of September 2007, associates reported that they voluntarily recycled the following:

- 675,538 lb of aluminum
- 282,476 lb of glass
- 5,953,357 lb of paper and cardboard
- 3,177,851 lb of plastic

They have also

- Walked, biked, and swam more than 1,109,421 total miles
- Cooked 368,779 healthy meals
- Quit smoking (nearly 20,000)
- Lost a total combined weight of 184,315 lb through PSPs
- Shared the PSP program with 375,824 of their friends, family, and community members outside of Wal-Mart

Due to the differences in each Wal-Mart community and the associates who call that region home, PSPs can be very different across the country. Stores might focus on wetlands conservation, recycling in the community, or education about climate change. And some stores are more engaged than others. Ultimately, our results have shown that our associates are motivated by PSPs when they save money, build friendships and support networks at work, and see the results of other associates' successful PSPs.

Tangible Results of PSP Program: Employees Making a Difference   Darryl Meyers, from the Burlington, NC, store noticed that the vending machines in the break rooms glowed with lights around the clock.

He suggested to Wal-Mart's energy division that they remove those lights to reduce energy consumption. Darryl not only made a suggestion that was good for the environment, but his idea also helped the company save nearly $1 million in electricity bills every year.

Another employee, Shonda Godley, who works in Wal-Mart's People Division at the Home Office in Bentonville, AR, is working to turn her family's farm into an educational, organic farm. Shonda's family farm traditionally produced wheat and maize, but now she would like to put the farm in a trust to a local college to be used as a training and learning facility for organic farming methods.

Challenges: Keeping it Fresh   As new associates join our company, and as others stay on, we have encountered a challenge of keeping the program informative and educational, while continuing to motivate and inspire long-time participants in the program. In 2008, we designed our PSP curriculum around feedback we received from our associates. By dividing the year-long curriculum into four topic-focused quarters, we were able to address some of the high-priority areas identified by associates. Our 2008 program educated associates on environmentally sustainable products carried in our stores, health and wellness, community engagement, and financial information and money-saving tips.

Next Steps: International Expansion   While PSP has never been officially introduced to our international markets, through word of mouth, many of our international associates are developing PSPs. We are hearing great stories from our associates in Canada, Japan, and China.

> *Sustainable SVNs: Aligning Business and Sustainability Goals*
> My challenge to you is to move sustainability to the front burner, if you don't already have it there, because it will be about your leadership and your future. We need to have 100 percent participation and we need to have full effectiveness in all of our efforts.
>
> **—Mike Duke, CEO of Wal-Mart Stores, Inc.**

Our associates are critical to our effort to become a more sustainable company. Since our sustainability initiatives affect so many business divisions, we established SVNs led by management associates within those divisions instead of simply creating a "Corporate Social Responsibility" division. When we launched our Sustainability 360 program, we wanted our associates to realize that this was a long-term program and part of our company-wide commitment to communities around the world. Though not designed as an educational program, the SVNs have been important in raising the awareness of environmental issues and identifying opportunities to improve our business and the environment simultaneously. Each SVN is responsible for developing initiatives that drive sustainability into specific business divisions and align with one of the overarching company sustainability goals.

Sustainable Value Networks

- Energy Goal: Greenhouse Gas Network, Alternative Fuels Network, Sustainable Buildings Network, Logistics and Fleet Network
- Waste Goal: Operations and Procurement Network (Waste), Packaging Network
- Products Goal: Food and Agriculture Network, Wood and Paper Network, Jewelry Network, Textiles Network, Chemicals Network and Electronics Network

Our SVNs include not only Wal-Mart associates but also representatives from nongovernment organizations (NGOs), supplier companies, academic institutions, government agencies, and other thought leaders who help us identify and execute meaningful changes in our business. Armed with the collective knowledge of each SVN, we can identify greater opportunities for improvement and develop innovative solutions in each business division. This collaborative approach has helped us in many instances, including our efforts to remove harmful chemicals from electronics and to bring sustainable fishing practices to our seafood suppliers around the world.

We involve salaried associates at all levels in the SVNs. In fact, the governance structure of the SVNs includes our executive leadership, top management, division leaders, and other associates at various levels throughout our company. Network activities are monitored and managed at a number of levels and are reported directly to our CEO.

Many of our managers were immediately engaged with the efforts through the SVNs, but to truly make sustainability live within our business we decided to make it every salaried associate's responsibility. Starting in 2008, we added a component to all of our salaried associates' annual reviews that evaluate their contribution to moving our company's sustainability goals forward.

We have found that our associates are bringing some of the best ideas to the table to make our company more sustainable.

Links to More Information

Wal-Mart sustainability: walmartstores.com/Sustainability/.

Wal-Mart fact sheets on PSPs and sustainability: walmartstores. com/FactsNews/FactSheets/#Sustainability.

Wal-Mart's Sustainability 2.0 video: www.walmart.com/catalog/ product.do?product_id=10237022.

MAPing a Route toward Sustainability

Case Study: Stonyfield
Company profile:
    Company: Stonyfield
    Industry: Food
    Headquarters: Londonderry, N.H.
    Number of Employees: 500
    Total Revenue: $340 million
Stonyfield's Story
    Nancy Hirshberg, Vice-president of Natural Resources

Synopsis: At Stonyfield, Inc. (Stonyfield) employee engagement and education on E&S issues begins with the hiring process. Then Stonyfield's Mission Action Plan (MAP) engages all employees in its sustainability mission through ongoing education and training and by linking long-term environmental impact goals to job performance measures of key personnel.

Why the Program Was Started: Engaging Employees in the Sustainability Mission One of the things I'm most proud of at Stonyfield is our Mission Action Plan (MAP) for engaging employees in our sustainability mission. Historically, Stonyfield has never screened potential employees for environmental literacy, so our employees are no more interested or knowledgeable about environmental issues than the general population. Through an assessment, we found that approximately 10% of our employees were managing 95% of the company's environmental impact. We began MAP in 2006 to engage more fully this key employee group in our sustainability mission. To our surprise, MAP evolved to engage all of our employees more fully in sustainability.

How the Program Works: Long-Term Goals Linked to Job Performance Measures MAP has truly transformed the company. MAP is built around Stonyfield's 11 primary areas of environmental impact, with cross-functional teams assigned to each impact area. The teams include transportation, facility, greenhouse gas emissions, milk production, sales, zero waste, green chemistry, water, sustainable packaging, ingredients, and SWOT (Stonyfielders Walking Our Talk). Unlike the other teams that focus on the company's areas of greatest environmental impact, SWOT focuses on high visibility but lower-impact issues like events and office supplies.

The MAP teams set long-term goals—such as zero waste—and complete yearly action plans for each goal. Our CEO, COO, and I must approve the plans. The team members also have a portion of their compensation linked to achieving an annual MAP objective. As a result of the MAP action plans, we now have company goals in place for our major areas of environmental burden, including facility energy, facility greenhouse gas emissions, percent renewable energy for our manufacturing facility, packaging and transportation. MAP team employees receive bonuses based on the achievement linked to these action plans. In addition, all capital improvement plans are reviewed and must be demonstrated to support the MAP goals. The result has been huge environmental savings such as elimination of solid waste and reduction of transportation greenhouse gas emissions.

The core group of employees involved in MAP felt that they had benefited so greatly professionally and personally from the program that they urged us to find a way to engage all of our employees in the process. In 2007, we held MAP trainings for every department. From production line workers to executives, everyone had an opportunity to learn about global environmental issues, with a focus on climate change. They learned about our company's impact on the environment and ways that they could help reduce environmental burdens at work and in their personal life. Also that year, we began a daylong orientation for all new employees on Stonyfield's mission, including MAP. We report monthly on MAP goals and progress in our in-house "moosletter." In 2008, one of the MAP goals—on facility energy use—became part of the bonus plan for all employees. We achieved the annual goal—reducing our energy use per ton of product by over 22%—and employees received a bonus. Currently, human resources are working to incorporate MAP goals into all employees' job descriptions.

Measuring Results: Gains in Key Areas   MAP has resulted in huge environmental savings. In 2007, we reduced transportation greenhouse gas emissions by 40%. We eliminated the equivalent of 18 tractor trailer loads of plastic.

We achieved a 28% reduction in facility energy use per ton product. And, 100% of Stonyfield products became organic.

In 2008, we achieved similar results by engaging employees in MAP. We reduced transportation greenhouse gas emissions by an additional 10% and energy use per ton of product by an additional 22%, compared to 2007! Our recycling rate is up by 13%, and our trash generated is down by 21% per ton of product. And, SWOT formed seven groups that are making improvements in a wide range of areas: office operations; clothing; premiums and gifts; travel and fuel; trash, compost, and recycling; food and caterers; office equipment and energy use; paper, printing, copying and faxing.

Challenges: Staying Focused   Despite the success we've achieved, we still face challenges. One of our biggest challenges is turnover and the need to continually educate new employees. It's also difficult getting people think "outside of the box"—to move beyond eco-efficiency to truly sustainable thinking. We also have to manage employee expectations. Many employees who have roles with very little impact want to be actively involved. We need to focus our efforts on the highest yielding projects—those with the greatest environmental and financial benefit. That involves very few people. So the challenge is supporting everyone else without diverting limited resources (time being key) away from key initiatives. There is never enough time or money.

Next Steps: Seeking Step Change   We're moving to the next level through training in innovation—thinking beyond eco-efficiency and more about products, processes, and practices that transform our company. We're looking now for the step changes that will move us to more sustainable business practices.

Through an assessment, we found that approximately 10% of our employees were managing 95% of the company's environmental impact. We began MAP in 2006 to engage more fully this key employee group in our sustainability mission. To our surprise, MAP evolved to engage all of our employees more fully in sustainability.

Key Lessons
- Tie E&S education to the company's mission and goals
- Focus on key impact areas and set improvement goals
- Make E&S relevant to job performance evaluation
- Assess E&S knowledge and motivation part of the hiring process

Additional information from the company case studies mentioned earlier indicates that while there is no "one-size-fits-all" approach to E&S education, engaging employees at every level of the company is essential to successful initiatives. But educational programs must compete for resources, so building a strong business case for an E&S education program can be as important as building the program itself. Several themes have emerged from the case studies as best practices for making the business case for E&S education:

- Corporate Strategy and Communications
  - Link E&S education initiatives to key business objectives and frame them in terms of management risks and opportunities.
  - Stress the shift in societal and stakeholder expectations. Sustainability is no longer just "nice to have" and

employees are an important resource for addressing and benefiting from this shift.

  - Take a top-down, bottom-up and sideways approach when engaging employees. A culture shift has to include everyone, not just those dedicated to sustainability.
- Creating and Managing Programs
  - Build momentum for the E&S actions by recognizing work that is already being done.
  - Create E&S education pilot programs that require few resources and measure the impacts of the pilot to build the case for a larger program.
  - Understand that each geographic region has its unique problems and opportunities.
  - Complement education with incentives (e.g., bonuses and awards) to improve environmental performance.
- Regularly report back to employees on how their E&S actions are making a difference.

While internal efforts are good, sometimes companies need to go externally for help with their programs. One of those companies providing support to an organization's efforts to engage employees is *GreenNurture* (http://greennurture.com [accessed October 29, 2014]).

> Your employees at all levels have valuable insights into how to make your organization more sustainable as they go about their duties. After all, it is their cumulative habits that have the greatest impact on how efficiently you use your resources. And by drawing them into the discussion, GreenNurture helps you utilize your human capital as capably as you use your financial capital!

As part of the GreenNurture program, employees receive the following:

- *Personal Home Page*—Your employees' personal space within the program, accessed via a username and password. It's their window into your campaign and the avenue by which they access all the tools they need to be active in it.
- *Forum*—The heart of the communication tool features a micro-blogging tool in which employees can submit ideas,

share best practices, and give feedback on issues you put forth to them.

- *Video Training*—Upload videos you want everyone to see and track who is participating.
- *Articles, News & Announcements, RSS feeds*—Raise awareness of important issues by selecting content to share with the entire organization, or even to smaller groups within the organization, and get their feedback on it so you can fine tune your sustainability campaign.
- *Recognition for Participation*—Change behavior with positive reinforcement, by giving points and badges to individuals for helping the organization's sustainability efforts.

Whether your mission/vision (strategic, sustainability, project) values to the "hub" of the Wheel have a strong internal connection, or need to have an external boost, one continually needs to "reflect" on the sustainability values of the organization. Without *reflecting* on those values, the true mission of sustainability may be lost.

4
The Tire

Introduction

In this chapter, we connect the enterprise's sustainability efforts, its ability to analyze and respond to sustainability threats, eliminate waste, and identify and develop opportunities for the enterprise to succeed in its sustainability efforts. It contains four surrounding dimensions: connect, detect, reject, and project. It is where "the rubber meets the road".

The connect dimension considers how well your enterprise does in terms of perception by the outside world of your commitment to sustainability, using measurements of the triple bottom line and corporate social responsibility. The detect dimension answers the question: "How well do we identify, analyze, and respond to sustainability-oriented threats?" The reject dimension captures the efforts of the enterprise to eliminate wastes from its processes. The project dimension identifies those opportunities that an enterprise can take advantage of if they are aware enough to do it.

Connect

> There are things known and there are things unknown, and in between are the doors of perception.
>
> –Aldous Huxley

This dimension of the Sustainability Wheel considers how well your enterprise does in terms of perception by the outside world of your commitment to sustainability, using measurements of the triple bottom line and corporate social responsibility.

The main question being asked here is this: *What do others think of our CSR and sustainability efforts, especially relative to others in our industry or practice area?*

The measurements here should be an intelligent composite of what is available for your particular area and will vary significantly by the following:

- The size of your enterprise
- The industry or practice area (IT, pharmaceutical, telecom, finance, government, consulting), and whether your particular focus is in service and/or manufacturing
- Number of employees
- Demographics of the employees
- Variety and type of customers

We provide some examples in the following but it will require you to put together a "net weighted result" within your context. This will be covered in Chapter 6—"Interpreting the Sustainability Wheel".

*DJSI**

From their web page:

The Dow Jones Sustainability™ Indices (DJSI) are maintained collaboratively by S&P Dow Jones Indices and RobecoSAM. Following a best-in-class approach, the indices measure the performance of the world's sustainability leaders. Companies are selected for the indices based on a comprehensive assessment of long-term economic, environmental and social criteria that account for general as well as industry-specific sustainability trends. Only firms that lead their industries based on this assessment are included in the indices. The indices are created and maintained according to a systematic methodology, allowing investors to appropriately benchmark sustainability-driven funds and derivatives over the long term.

The family includes global and regional broad market indices, sub-indices excluding alcohol, gambling, tobacco, armaments and firearms

---

* http://www.sustainability-indices.com/.

and/or adult entertainment, and global and regional blue-chip indices. The most widely referenced Dow Jones Sustainability™ Indices are listed in the following. For additional information on the Dow Jones Sustainability™ Indices, including a full suite of index literature, visit the official index website.*

Table 4.1 shows the breadth and depth of these indices.

Table 4.1   Significant Breadth to the Dow Jones Sustainability Index

Dow Jones Sustainability™ Asia/Pacific Index
Dow Jones Sustainability Australia Index
Dow Jones Sustainability Emerging Markets Index
Dow Jones Sustainability Europe Index
Dow Jones Sustainability Eurozone Index
Dow Jones Sustainability Korea Index
Dow Jones Sustainability North America Index
Dow Jones Sustainability United States Index
Dow Jones Sustainability World Developed Index
Dow Jones Sustainability World ex Switzerland Index
Dow Jones Sustainability World Index
Dow Jones Sustainability World 80 Index
Dow Jones Sustainability World ex US 80 Index
Dow Jones Sustainability North America 40 Index
Dow Jones Sustainability United States 40 Index
Dow Jones Sustainability Asia/Pacific 40 Index
Dow Jones Sustainability Europe 40 Index
Dow Jones Sustainability Eurozone 40 Index
Dow Jones Sustainability Canada Select 25 Index
Dow Jones Sustainability Japan 40 Index
Dow Jones Sustainability Asia/Pacific Index ex Alcohol, Tobacco, Gambling,
    Armaments & Firearms
Dow Jones Sustainability Europe Index ex Alcohol, Tobacco, Gambling, Armaments &
    Firearms and Adult Entertainment
Dow Jones Sustainability North America Index ex Alcohol, Tobacco, Gambling,
    Armaments & Firearms
Dow Jones Sustainability World Enlarged Index ex Alcohol, Tobacco, Gambling,
    Armaments & Firearms and Adult Entertainment
Dow Jones Sustainability World Index ex Alcohol, Tobacco, Gambling, Armaments &
    Firearms
Dow Jones Sustainability Korea 20 Index

---

* http://www.sustainability-indices.com/.

Claremont-McKenna's Roberts Environmental Center Pacific
Sustainability Index (2006 through 2013)

The Pacific Sustainability Index (PSI)* is an index generated by the results to questionnaires to analyze the quality of the sustainability reporting. There are actually two questionnaires: (1) a base questionnaire for reports across sectors and (2) a sector-specific questionnaire for companies within the same sector. The methodology used by Roberts Environmental Center (REC) involves analysts downloading relevant English web pages from the main corporate websites for analysis. They exclude data independently stored outside the main corporate website. If a corporate subsidiary has its own sustainability reporting, partial credit is given to the parent company when a direct link is provided in the main corporate website. Analysts filled out a PSI scoring sheet and tracked the coverage and depths of different sustainability issues mentioned in all online materials.

Then, analysts enter their scoring results into a PSI database. The PSI database determines scores and publishes them on the center's website. The sector reports provide an in-depth analysis on sustainability reporting of the largest, but not more than 30, companies of the sector, as listed in the latest Fortune Global 500 and 1000 lists. Prior to publishing sector reports, the REC notified and encouraged companies analyzed to provide feedback and additional new online materials, which often improved their scores.

In Figure 4.1, you can see the ranking of companies' sustainability performance in PSI's most recent report—a 161-page analysis of the telecom sector. Each company gets a grade and score. Further, as a way to gain some insight into the analysis, in Figure 4.2, you can see the percentage of companies in their study who were addressing a wide variety of PSI's environmental topics.

It turns out that Claremont-McKenna is no longer performing the PSI. Contacted for comment, Dr. William Ascher of the University said, "We are no longer doing the PSI, in light of the fact that most large corporations now do a quite good job of mentioning the environmental and other CSR dimensions that the PSI has covered (whether

---

* Example here: https://www.claremontmckenna.edu/roberts-environmental-center/
wp-content/uploads/2014/02/2005FoodReport.pdf.

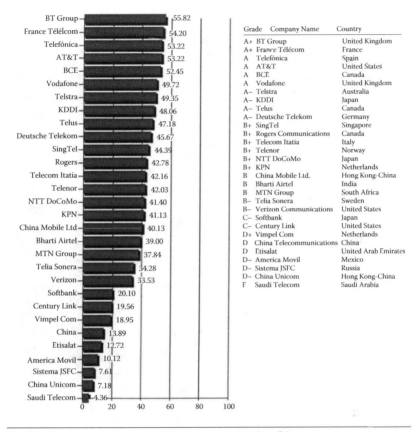

Grade	Company Name	Country
A+	BT Group	United Kingdom
A+	France Télécom	France
A	Telefónica	Spain
A	AT&T	United States
A	BCE	Canada
A	Vodafone	United Kingdom
A−	Telstra	Australia
A−	KDDI	Japan
A−	Telus	Canada
A−	Deutsche Telekom	Germany
B+	SingTel	Singapore
B+	Rogers Communications	Canada
B+	Telecom Itatia	Italy
B+	Telenor	Norway
B+	NTT DoCoMo	Japan
B+	KPN	Netherlands
B	China Mobile Ltd.	Hong Kong-China
B	Bharti Airtel	India
B	MTN Group	South Africa
B−	Telia Sonera	Sweden
B−	Verizon Communications	United States
C−	Softbank	Japan
C−	Century Link	United States
D+	Vimpel Com	Netherlands
D	China Telecommunications	China
D	Etisalat	United Arab Emirates
D−	America Movil	Mexico
D−	Sistema JSFC	Russia
D−	China Unicom	Hong Kong-China
F	Saudi Telecom	Saudi Arabia

Figure 4.1 The Pacific Sustainability Index—an example from Telecom.

they are doing a good job in environmental protection is another, much thornier question)."

Still, the analysis they've done in the past is worth looking at, both for methodology and for results, if you'd like to understand how such an analysis was done for over 10 years.

GISR* (Global Initiative for Sustainability Ratings)

From their website:

> Launched in June 2011 as a joint project of Ceres and Tellus Institute, the Global Initiative for Sustainability Ratings (GISR) is a new participant in the family of initiatives aimed at making financial markets

---

* http://ratesustainability.org/about/.

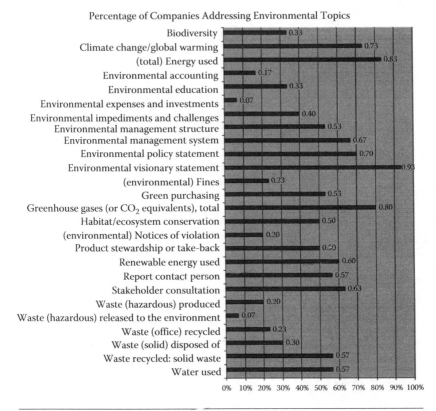

Figure 4.2   Summary of indices for global initiative for sustainability ratings.

agents of, rather than impediments to, achieving the Post Rio+20 global sustainability agenda. As a global, multi-stakeholder initiative, its vision is to transform the definition of corporate value in the 21st century such that markets reward the preservation and enhancement of all forms of capital—human, intellectual, natural, social and financial.

GISR's mission is to design and steward a global sustainability (i.e., Environmental, Social, and Governance—ESG) ratings standard to expand and accelerate the contribution of business and other organizations worldwide to sustainable development. GISR will not rate companies. Instead, it will accredit other sustainability ratings, rankings or indices to apply its standard for measuring excellence in sustainability performance.

Underlying GISR's vision and mission is the core premise that a globalizing and resource constrained world will be well served by convergence around a generally accepted definition of what constitutes corporate sustainability excellence. Just as such norms have evolved the fields of human rights, labor practices, and sustainable forestry, so too should a common understanding of the core elements that define excellence in sustainability performance. GISR believes that achieving this goal through an inclusive, adaptable process will serve as a powerful driver in moving companies and markets alike toward continually higher levels of contribution to long-term, global social and ecological well-being.

CSRHub.com

From their web page*:

CSRHub is a web based tool that provides access to employee, environmental, community and governance ratings on most major companies in North America, Europe and Asia. We are the first company to combine data from nine of the premier socially responsible investment (SRI) analysis firms (also known as Environment, Social, Governance - ESG), and over 265 nongovernmental organizations (NGOs), government agencies, news feeds, social networking groups, smaller for-profit organizations, and publishers. Our proprietary tools combine more than 60 million pieces of data on sustainability and CSR performance into a consistent set of ratings. We then allow users to personalize these ratings, share them, and add their own views on companies. Our site enables users to learn about and compare company sustainability and CSR behavior. We provide some ratings information for free and additional information to fee-paying subscribers. We also sell custom reports from most of our licensed ESG sources.

Sustainability Leadership Report†: Brandlogic and CRD Analytics

Sustainability Reality Score (SRS), created from

- 175 metrics for rating companies
- 5 key performance indicators per ESG (Economics, Social, Governance) dimension
- 1200 rated corporations

---

* http://www.csrhub.com/content/about-csrhub/.
† http://www.sustainabilityleadershipreport.com/downloads/2012Sustainability_leadership_report.pdf.

Sustainability Perception Score (SPS), created from

- 16,000+ company ratings
- 2,400 respondents from 3 "most attentive" stakeholder segments
- 100 prominent global corporations covering 9 of the 10 global industry categories

This yielded the following:

Leaders: Those who excel in both real and perceived performance
Promoters: Those with relatively high perceived performance, but relatively low real performance
Challengers: Those with good real performance but relatively low perception ratings
Laggards: Companies that trail on both dimensions

The company also shows the amount of gap between SPS and SRS, indicated by the size of the enterprise's "dot" on the IQ matrix.

These 100 prominent global corporations then show up in 1 of the 4 categories earlier, which are represented by the 4 quadrants of the matrix. In the following, we show the foundational IQ matrix and a snapshot of the 2012 IQ report (Figures 4.3 and 4.4).

When asked about the SustainabilityIQ assessment and how it intersects with this book, Hampton Birdwell, CEO and Managing Partner of Tenet Partners said,

> Ultimately, the alignment of brand perceptions and operations performance on sustainability issues is critical for managing corporate reputation and brand image. When done well, corporate leaders and managers have the opportunity to provide compelling communications to stakeholders that support the corporate vision and accurately reflect the activities of an organization as it addresses the sustainability of the enterprise. It's vital that organizations get this right to avoid the pitfalls of over stating their efforts, thus introducing reputation risk to the company.

You can see from Birdwell's quote how this connect dimension will interact and overlap somewhat with the detect dimension.

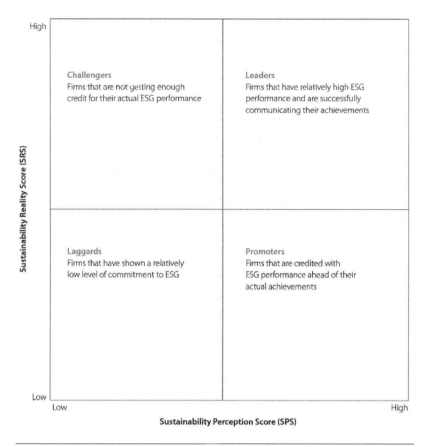

High

Challengers
Firms that are not getting enough
credit for their actual ESG performance

Leaders
Firms that have relatively high ESG
performance and are successfully
communicating their achievements

Laggards
Firms that have shown a relatively
low level of commitment to ESG

Promoters
Firms that are credited with
ESG performance ahead of their
actual achievements

Low

Low   High

Sustainability Reality Score (SRS)

Sustainability Perception Score (SPS)

Figure 4.3 The sustainability IQ matrix—sustainability reality plotted against sustainability perception.

*Newsweek Green Rankings**

From their website:

> Newsweek Green Rankings is one of the world's foremost corporate environmental rankings. The project ranks the 500 largest publicly-traded companies in the United States (the U.S. 500) and the 500 largest publicly-traded companies globally (the Global 500) on overall environmental performance.

---

* http://www.newsweek.com/green-2014.

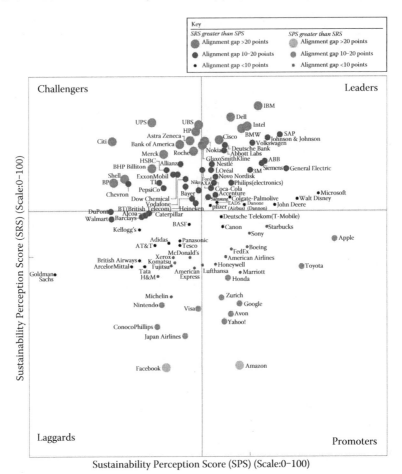

Figure 4.4 Example company results in the substitutability IQ format.

This is determined using criteria from the Table 4.2 with accompanying weight:

KPI measurements for Newsweek Green 500 are based on six key principles.

Transparency: The precise methodology of the ranking and the results of the process are fully disclosed.

Objectivity: Eligible companies will only be assessed using quantitative data and performance indicators.

Public data: Only data points that are part of the public domain are used.

Table 4.2 How the Newsweek 500 Key Performance Indicators are Used

NEWSWEEK 500 KEY PERFORMANCE INDICATOR	NOTES ON THE INDICATOR	WEIGHT (%)
Energy Productivity	Energy Productivity defined as Revenue ($U.S.)/Total Energy Consumption (in gigajoules)	15
GHG Productivity	GHG Productivity defined as Revenue ($U.S.)/Total Greenhouse gas (GHG) Emissions (CO2e)	15
Combined Water Productivity	Water Productivity is defined as Revenue ($U.S.)/Total water (in cubic meters).	15
Waste Productivity	Waste Productivity is defined as Revenue ($U.S.)/[Total waste generated (metric tonnes) − waste recycled/reused (tonnes)].	15
Reputation − Green Sanctions	The total amount of environmental fines, penalties and settlements paid in the year 2012 or deemed in the year 2012 to be payable by the company, irrespective of when the actual cash flow occurs are divided by the company's total revenue for 2012. The ratio is percent ranked against that of all Industry Group peers. Only the fines, penalties and settlement amounts that are definitive (i.e., the company has no other recourse but to pay) are considered.	15
Sustainability Pay Link	A mechanism to link the remuneration of any member of a company's senior executive team with the achievement of environmental performance targets. The existence of such a link is awarded a score of 100%. A score of 0% is attributed if there is no such mechanism in place.	15
Sustainability Board Committee	The existence of a committee at the Board of Directors level whose mandate is related to the sustainability of the company, including but not limited to environmental matters. A score of 100% is awarded if such a committee exists, and a score of 0% is given in cases where such a committee is absent.	5
Audited Environmental Metrics	The company provides evidence that the latest reported environmental metrics are audited by a third party. A score of 100% is awarded if such an audit has been performed, and a score of 0% is given in cases where such an audit was not performed.	5

Comparability: Companies are compared against their industry group peers based on performance indicators for which the underlying data are reasonably well disclosed by their industry group globally.

Engagement: Companies eligible for the ranking will be informed prior to the ranking, so as to have an opportunity to ensure the necessary data are made available publicly.

Stakeholders: Stakeholder feedback is actively solicited throughout the project. A panel of experts, consisting of eight leading sustainability practitioners, reviewed and commented on all aspects of the Newsweek Green Rankings methodology.

ClimateCounts.org

From their website*:

Climate Counts is a collaborative effort to bring consumers and companies together to address solutions around global climate change.

We score the world's largest companies on their climate impact to spur corporate climate responsibility and conscious consumption. Our goal is to motivate deeper awareness among consumers—that the issue of climate change demands their attention, and that they have the power to support companies that take climate change seriously and avoid those that don't.

When consumers take action and raise their voices on issues that matter to them, businesses pay attention.

We have no interest in doom-and-gloom environmental reporting and instead believe that positive change starts with a hopeful outlook that real change is possible and that the relationship between companies and consumers can become more substantive and constructive.

Resources:
Still don't believe in climate change? We don't claim to be scientists, so we rely on information from the following websites that comes from people who know what they're doing:

- National Aeronautics and Space Administration (NASA)— Global Climate Change website.
- National Geographic—Greenhouse Effect interactive website.
- Intergovernmental Panel on Climate Change (IPCC).
- Real Climate—Climate Science from Climate Scientists.
- Climate Counts board member and University of New Hampshire Research Associate Professor Dr. Cameron Wake website.

---

* http://climatecounts.org/about.php.

Examples of actual Climate Counts reports retrieved from the Internet in late 2014 are given later. We've chosen companies at random that appear at the two ends of the spectrum. Note that the report shows not only the results but also the change in results from last year to the current year (Figures 4.5 and 4.6).

In what follows are the ClimateCounts.org criteria by topics (REVIEW, REDUCE, POLICY STANCE, and REPORT), with the subtotals shown at the top. The entire scorecard adds to 100 points. By looking at the "points" column on the right, you can get an idea of the weight that ClimateCounts is giving to the subcategories (Table 4.3).

Company Scorecard

amazon.com   Amazon.com

Stuck   9

STUCK - A choice to avoid for the climate-conscious consumer. This company is not yet taking meaningful action on climate change.

Let companies know you care:

@amazon

Change from previous year's score: -2

Review: 0/22 points. Amazon.com has not made efforts to measure its companywide impact on global warming (i.e., its greenhouse gas emissions or climate footprint).

Reduce: 8/56 points. Amazon.com has taken basic steps to reduce the company's energy use.

Policy Stance: 0/10 points. Amazon.com has provided no public information that supports public policy that addresses climate change.

Report: 1/12 points. Amazon.com has made some public information available on its efforts to address global warming.

CLICK HERE TO TELL THIS COMPANY YOU THINK CLIMATE COUNTS!

Figure 4.5   Example climate counts scorecard for Amazon.

Company Scorecard

 Unilever

  91

Soaring

SOARING - The best Climate Counts choice. These companies are demonstrating exceptional leadership on climate change, but realize there is always room for improvement.

Let companies know you care:

@Unilever
@Bertolli
@CountryCrock
@Hellmanns
@Lipton
@Slimfast
@Breyers
@Klondikebar
@WishBoneBrand
@Dove
@AXE
@VaselineBrand

Change from previous year's score: +3

Review: 21/22 points. Unilever annually measures its companywide impact on global warming.

Reduce: 48/56 points. Unilever has established clear goals to reduce the company's energy use and has achieved reductions in its impact on global warming (i.e., its greenhouse gas emissions or climate footprint).

Policy Stance: 10/10 points. Unilever has distinguished itself by strongly advocating for comprehensive public policy that addresses climate change and would lead to market-wide reduction in greenhouse gas emissions and the growth of renewable energy.

Report: 12/12 points. Unilever has made public information available on its companywide efforts to address global warming.

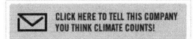 CLICK HERE TO TELL THIS COMPANY YOU THINK CLIMATE COUNTS!

Figure 4.6 Example climate counts scorecard for unilever.

Global 100

From their website*:

Determining which companies are "sustainable" and which are not is a challenging enterprise. Not only is there no single, universally accepted definition of "corporate sustainability," publicly traded companies are exceedingly complex institutions, often spanning multiple geographies and industrial sectors.

Against this backdrop, our approach is simple—we unpackage "corporate sustainability" into its component parts, and stick to the numbers.

---

* http://www.corporateknights.com/reports/global-100/.

Table 4.3 Weighted Elements of the Global 100

Review	**22**
GHG emissions inventory completed?	5
Rough calculations or standard protocol/calculator?	3
Are Kyoto gases besides CO_2 included?	2
Are indirect emissions accounted for (e.g., supply chain, travel, commuting, use/disposal of products/services, and investment)?	4
Is there external, qualified third-party verification of emissions data, reductions, and reporting (where applicable)?	4
Is the inventory an ongoing, regular process accounting for multiple years?	4
Review	**56**
Has the company achieved emissions reductions?	10
Has the company taken steps toward achieving reduction target? (interim progress on reduction)	8
Magnitude of reduction goal	5
Have a management plan and organizational structure been established for climate?	5
Has the company achieved verified reductions to date (prior to current goal setting)?	5
Has a clear goal been set?	4
Absolute or intensity-based reductions?	4
Has the company made successful efforts to reduce GHG impacts associated with the use of its products/services?	4
Does the company work to educate its employees, trade association, and/or customers on how they can reduce individual GHG emissions (through direct education programs, incentives, or philanthropic projects)?	4
Strength of baseline year used for the reduction goal? (keeping in mind changes in company's size/composition)	3
Is there top-level support for climate change action?	2
Does the company require suppliers to take climate change action or give preference to those that do?	2
Policy Stance	**10**
Does the company support public policy that could require mandatory climate change action by business?	10
Does the company oppose public policy on climate change that could require mandatory action by business, or has it made efforts to undermine climate change action?	−10
Report	**12**
Is the company publicly reporting on emissions, risks, and actions? How is information disclosed? Company based (e.g., on their website or annual report) or through a credible third-party program (CDP, GRI, etc.)?	10
Are emissions broken out by facility, business unit, country of operations, or other meaningful subsegments?	2

Qualifying Global 100 companies are scored on a percent rank basis against their global industry peers on a list of twelve quantitative key performance indicators that run the gamut from energy and water use, to employee compensation and corporate tax strategy.

Because the indicators are quantitative and clearly-defined, the results of the Global 100 are objective and replicable. The Global 100 follows a rules-based construction methodology and is more akin to a financial index than many other "sustainability" indices.

Here are the criteria used by the Global 100:

- Energy Productivity
- Carbon Productivity
- Water Productivity
- Waste Productivity
- Innovation Capacity
- Percentage Tax Paid
- CEO to Average Worker Pay
- Pension Fund Status
- Safety Performance
- Employee Turnover
- Leadership Diversity
- Clean Capitalism Pay Link

As an example description of one of the criteria, Energy Productivity (all are available at their website*) provides some insight as to how they are measured:

In just about every jurisdiction on Earth, energy costs are rising. Prices are also becoming much more volatile, making it more difficult for companies to manage their energy strategy. This metric looks at how much revenue companies can squeeze out of every unit of energy they use, and shows which companies are best able to adapt to our changing energy future.

Equation: Revenue ($US)/Energy use (Gigajoules)

---

* http://global100.org/key-performance-indicators/.

Understanding your reflect dimension can be very helpful as we move to the next dimension, detect. Why? Well, look back at our opening quote: "There are things known and there are things unknown, and in between are the doors of perception."—Aldous Huxley. What we've done here is to gain more knowledge (that is to bring more into the "known" domain, regarding what stakeholders think about our work, our results, in the area of sustainability. However, certainly, there is more to discover—in this dimension and others—in terms of what those unknowns translate to in terms of threat. And that is what the detect dimension is all about.

Detect

> Wherever there is a danger, there lurks an opportunity; wherever there is opportunity, there lurks danger. The two are inseparable.

> **–Earl Nightingale**

This chapter reviews the detect dimension. It answers the question: "How well do we identify, analyze, and respond to sustainability-oriented threats?"

Note: Recalling the dual nature of risk promoted by the Project Management Institute in its *PMBOK® Guide*, it's important to understand that risk has two components: negative risk (threats) and positive risk (opportunities). This chapter focuses on threats. For a holistic treatment of CSR-related risk, we highly recommend that you balance your reading of this book by also reviewing the "Project" section, which focuses on opportunities.

We begin with the same recommendation we give to all of our clients and to our students of project management when it comes to identifying threats: start with the stakeholders. A broad and deep identification of stakeholders is important to a broad and deep identification of risk (especially negative risk, or threats). We'll provide more on threat identification in general a bit later in the chapter, but let's start with an excellent article* on stakeholder management with a sustainability perspective.

The Sustainability Side of Project Stakeholder Management

By Gratien Gasaba

(*Reprinted here as originally posted with permission from the author.*) Practitioners in development projects know that stakeholder management alongside with communication management is not only important but also mandatory for project success. Some project stakeholders may be so influential that they can do whatever they want. The art of project management requires sometimes "getting elephants to dance to your song." In a sustainability-driven management, the project anthem to be sung by everyone is the survival of results after the intervention is closed. The rest of this discussion attempts to show how the stakeholder management can be done in a way that promotes sustainable development.

SUSTAINABILITY STAKEHOLDERS' IDENTIFICATION AND ANALYSIS

Before the implementation of a project, it is always recommended to identify and analyze stakeholders with a focus on sustainability aspects. Once stakeholders are identified, the next step will be to analyze them. The output of the analysis exercise is a classification of all stakeholders in categories according to their support or

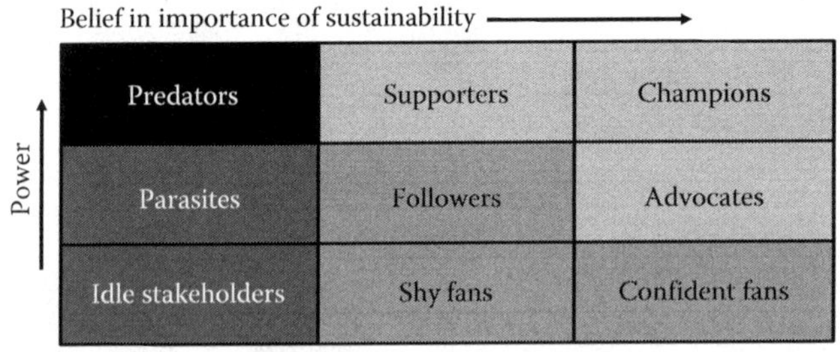

Belief in importance of sustainability ⟶

Predators	Supporters	Champions
Parasites	Followers	Advocates
Idle stakeholders	Shy fans	Confident fans

Power ↑

Figure 4.7 Power level vs. sustainability support level. (Adapted with permission of Gratien Gasaba.) Source was PMhut.com, http://www.pmhut.com/sustainability-side-of-project-stakeholder-management, accessed on November 20, 2014.

---

* http://www.pmhut.com/sustainability-side-of-project-stakeholder-management.

interest in lasting solutions and the level of power they have to do what they want.

There are several ways to analyze stakeholders focusing on the sustainability aspects of the project. For instance, the analysis can use a power and sustainability grid as summarized in Figure 4.7.

Predators

These are powerful people who don't care about lasting solutions which they most often perceive as threats to their personal interests. In their efforts to safeguard their own interests, they seriously harm project sustainability. These personal interests are often related to their professional position, private business, or even political aspirations. Not only sustainability predators have means and power to destroy sustainability efforts, but also they actively attack whoever advocates or support the survival of project decisions. Predators may be actors inside or outside the project organizational environment. It is note-worthy to mention that, most dangerous predators are those within the organizational environment, who use hidden strategies to attack lasting solutions and get them aborted before they are implemented.

Parasites

These are people with a portion of power and whose interests are against sustainability. They don't openly oppose the sustainability plan, but they never support it. Additionally, some of their personal inter-ests, actions, and behavior may harm the survival of the project results. They benefit from the project's immediate results, but never take action to protect and keep those results for long time. Some business men and women, functional managers, political actors, and project team mem-bers may fall in this category when they don't resolve their respective conflicts of interest during sustainability-related decision making.

Idle Stakeholders

These are stakeholders with very low level of power and with almost no interest or information on sustainability issues. They behave like newborns. They do nothing that can promote sustainability or harm it.

They can also be called wrong sustainability stakeholders because they don't have open interest in lasting solutions or influence on the survival of project results after its closure.

Shy Fans

Shy fans of sustainability are those project technicians with relevant skills and knowledge but who cannot actively defend project long-term deliverables though they are not against them. They have almost no influence in the project, not because there is no room for influence, but as a result of lack of interest or motivation. If they are motivated, they can become sustainability followers or confident fans. Shy fans can also be end users of project results who are interested in lasting solutions, but with little or no information on how to defend their needs.

Shy fans may also be small-scale businesses or individual novice entrepreneurs who, at the same time, depend on the predators though they have real interest in lasting solutions. They don't speak their mind because they fear predators.

Confident Fans

They have high-level support for sustainability, but their level of influence is almost none. They openly claim to be in favor of project lasting results. The word sustainability is always in their speech and mobilizes all stakeholders in favor of the survival of project results after closure. Though they lack influence due to limited information and other means, confident fans tirelessly raise their voices in defense of long-term solutions but they are unable to show how to implement them. The enthusiastic attitude of confident fans results from their liberal nature that sets their minds free to express themselves in favor of sustainability. The only barrier to confident fans is the low level of hard power, which, to a certain degree, is a result of low economic, social, political, or intellectual level.

Followers

They have medium power. They are middle managers with some useful technical skills who are interested in sustainability of the project.

Their level of support for lasting solutions is balanced by their wish to obey their bosses who may not be interested in sustainability. They tend to follow the balance of power in the project environment.

Followers may also be middle-scale businesses or individual entrepreneurs watching the power game around sustainability between very active stakeholders (for instance between predators and champions). They are ready to support and advocate for sustainability if certain conditions are met. They are also called opportunists because they support sustainability only when conditions pertaining to their own interests are met.

Supporters

These are people with high level of power and a lot of activities, with medium support to sustainability. Champions and advocates use high-level influence of supporters to promote and protect lasting solutions and to thwart the predators' actions and plans. The behavior of supporters is carefully monitored by champions and advocates, to ensure they don't fall in the hands of predators. The alignment of supporters is a very determining factor of the outcome of the game around sustainability.

Advocates

These are the people with high-level support of the project sustainability and with some portion of power. They have relevant knowledge and skills about sustainability. They have much interest to defend lasting solutions and they do all they can to lobby other stakeholders including high project decision makers. Their heart is on the sustainability and all their energy is allocated to it. Their only barrier is possible action by predators who are more powerful and ready for all against viable solutions. Champions, strongly rely on advocates' support to crush predators and parasites. It is this alliance between champions and advocates that makes a difference in the campaign for sustainability.

Champions

The sustainability champion is an exceptional individual person who devotes himself or herself for lasting solutions in a challenging and consistent way. He/she is one of the project stakeholders. The sustainability championship role may also be played by an organization. An organization that has maturity in a given sector and that has integrated sustainability issues in its operations and decision-making procedures may play a lead role in mobilizing sister organizations toward lasting solutions.

Sustainability champions are good at advocating and lobbying. They always find the way to get support from both operational and strategic levels. They are skilled in mobilizing people around sustainability dimensions of a project. Because of their high level of support in lasting solutions and influence in the organization, champions play a critical role in the sustainability assurance.

Conclusion

The old thinking, that sustainability issues are mostly technical, with little governance dimension, no longer holds. Historical evidence has shown that the human factor, which is by essence social and political, plays a leading role in shaping sustainability. Apart from natural action, the rest of what happen in our environment results from what people do. Since what people do shapes sustainable development, there is a possibility to orient sustainability through human actions, and stakeholder management has its place here.

As Gratien describes earlier, you can see the tight connection between stakeholder identification and risk identification—threat identification in particular. We recommend using the stakeholder identification process hand in hand with the risk identification process and using the specific advice given in the article about how "sustainability stakeholders" deserve their own special focus when you identify stakeholders *and* risk. This will help you assure that Corporate Social Responsibility (CSR) risks are being detected early on—when you can do something about them with thoughtfulness, foresight, and proper planning—instead of duct tape, twine, and superglue.

Any good project manager will tell you that when you are looking for risks (threats in particular) you look for risk triggers. A trigger is something that tells you definitively that the risk has occurred or that it is about to occur—or at least that conditions are ripe for it to occur. We often use the example of a forest ranger in their lookout tower.

Of course, smoke is a trigger for the threat of fire, but so is a series of 10 consecutive dry days.

One of the ways we can look for risk triggers in the area of sustainability is in the demonstrated behaviors of our project and program teams. One trigger behavior to note is related to motivation and before we get there, we're going to take a step back and look at motivation and drive.

An author of note in this area is Dan Pink. In his book, *Drive: The Surprising Truth About What Motivates Us*,[*] Pink talks about Autonomy, Mastery, and Purpose. He sums it up this way: When it comes to motivation, there's a gap between what science knows and what business does. Our current business operating system—which is built around external, carrot-and-stick motivators—doesn't work and often does harm. We need an upgrade. And the science shows the way. This new approach has three essential elements: (1) *Autonomy* — the desire to direct our own lives. (2) *Mastery*—the urge to get better and better at something that matters. (3) *Purpose*—the yearning to do what we do in the service of something larger than ourselves.
In his book, *Purpose*,[†] Nikos Mourkogiannis says,

> I believe that Purpose—not money, not status—is what people most want from work. Make no mistake: They want compensation; some want an ego-affirming title. Even more, though, they want their lives to mean something, they want their lives to have a reason.

We're going to focus on the Purpose piece here. The purpose motive is about having a job—or in the case of a project, an objective or task—that is about something meaningful, something lasting, something generally meant to improve the lot for others. Of course, economic success is important, but if we look at the mission statements of organizations (see Chapter 1) we know that today more and more companies recognize that there is a larger, more holistic, longer-lasting definition of success. So, people tend to be motivated to purpose,

[*] D.H. Pink (2009) *Drive: The Surprising Truth About What Motivates Us*, Riverhead Books, New York.
[†] N. Mourkogiannis (2006) *Purpose: The Starting Point of Great Companies*, Palgrave MacMillan, New York.

leaders of companies recognize and promote this, but sometimes at the project level, the team can tend to suboptimize toward profit. But what Pink points out is this: "When the profit motive gets unmoored from the purpose motive, bad things happen." Bad things include the following:

- Poor quality, in terms of requirements being gathered with our own rationale and no connection to the customer's needs
- Poor customer service (the customer becomes a secondary concern to, for example, operational costs)
- Intra-organizational conflict

So, coming back to risk triggers, if we notice that a project team has become unmoored, that would tell us that perhaps they are suboptimizing in terms of overall success. And in turn, a way to notice this, ironically, is by looking at the project teams' risk registers. Risk registers—like the stakeholder registers we discussed earlier—need to be broad and deep in nature. That is, they should cover all sorts of threats—regulatory, economic, competitive, technical, materials, resources, as well as social and ecological risks. One doesn't have to go too far back in history—or to a minor incident—for an example.*

The Deepwater Horizon oil spill (also known as the Gulf of Mexico Oil Spill or the BP Oil Spill) is the largest marine oil spill in history, and was caused by an explosion on the Deepwater Horizon offshore oil platform about 50 miles southeast of the Mississippi River delta on April 20, 2010 (28.74°N, 88.39°W). Most of the 126 workers on the platform were safely evacuated, and a search and rescue operation began for 11 missing workers. The Deepwater Horizon sank in about 5,000 feet (1,500 m) of water on April 22, 2010. On April 23 the U.S. Coast Guard suspended the search for missing workers who are all presumed dead. After a series of failed efforts to plug the leak, BP said on July 15 that it had capped the well, stopping the flow of oil into the Gulf of Mexico for the first time in 86 days.

The oil slick produced by the Deepwater Horizon oil spill covered as much 28,958 miles² (75,000 km²), an area about the size of

---

* Drawn mainly from http://www.eoearth.org/view/article/161185/.

South Carolina, with the extent and location of the slick changing from day-to-day depending on weather conditions. By the first week in June, oil had come ashore in Louisiana, Mississippi, Alabama, and Florida, with significant wildlife fatalities in Louisiana. In the weeks following the accident, scientists discovered enormous oil plumes in the deep waters of the Gulf of Mexico, raising concerns about ecological harm far below the surface that would be difficult to assess.

The surface slick threatened the ecosystems and the economy of the entire Gulf Coast region. The U.S. Fish and Wildlife Service reported that up to 32 National Wildlife Refuges were potentially affected by the spill. Concerns were raised about the environmental impacts of chemicals known as dispersants that have been used to dissipate the oil slick. By June 2, 2010, the National Oceanic and Atmospheric Administration (NOAA) had banned fishing in about 36% of federal waters, or 86,895 miles^2 (229,270 km^2) of the Gulf.

By June 9, BP stock had lost close to half its value, more than $82 billion, in the 7 weeks since the spill started, although the stock rebounded somewhat on the fall of 2010. According to BP, the cost of the response to September 29 amounted to approximately $11.2 billion, including the cost of the spill response, containment, relief well drilling, static kill and cementing, grants to the Gulf States, claims paid, and federal costs (Figure 4.8).

Bloomberg News September, 2014: "BP Plc acted with gross negligence in setting off the biggest offshore oil spill in U.S. history, a federal judge ruled, handing down a long-awaited decision that may force the energy company to pay billions of dollars more for the 2010 Gulf of Mexico disaster."

The financial costs continue to pile up, with total fines estimated in 2014 to be more than $50 billion. And the costs to the company are of course greater than financial, they involve brand reputation. In fact even though for some people this is "old news," the story continues to make news in 2014, with the company now facing gross negligence charges. And of course, we cannot forget that there was loss of life and the ongoing cost to the environment which is still being assessed today.

We often point out that the risk register for Macondo (BP's well, within the Deepwater Horizon platform owned by Transocean) included precisely zero safety or environmental risks. In fact, we

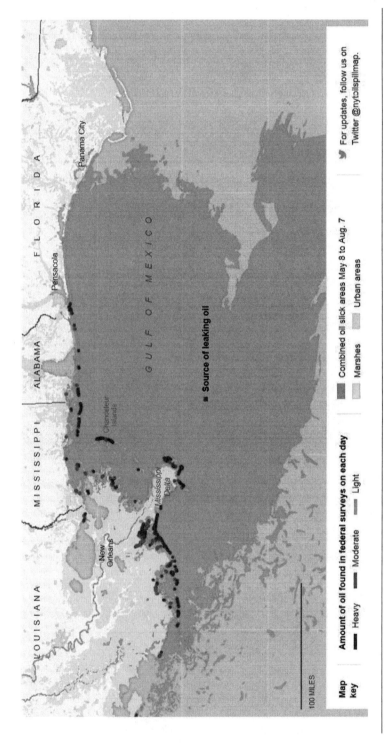

Figure 4.8   Oil from BP's Macondo Well, as discovered by US federal surveys. Credit, NY Times, site http://www.nytimes.com/interactive/2010/05/01/us/20100501-oil-spill-tracker.html, accessed on September 9, 2014 (Map source: *New York Times.*)

recently obtained the Microsoft Excel version of the spreadsheet and we were further surprised that the "drop down menu" for risk impacts didn't even provide the possibility for adding safety or environmental risks (see Figure 4.9).

All of the threats identified had to do with operations and efficiency. This is public information because of the Federal U.S. investigation into the disaster. Part of the reason could be that even though at the high level BP was making statements about its commitment to safety and the environment, other leaders at BP had just introduced an incentive plan for managers which rewarded them significantly for efficiency and operational metrics but not at all for any safety or sustainability measures. That's a case of profit motive being unmoored from the purpose motive.

So, our tip for the portfolio managers and program managers is to audit the risk registers of your projects to verify that the purpose motive is present; that the "mooring" between the enterprise's purpose is solid in your projects.

One other thought we'd like to revisit here, and it actually deals with both the threats and opportunities. Looking at Figure 4.10, which adapts the "Swiss cheese" model originally proposed by Dante Orlandella and James T. Reason,* we show how the understanding of Change Quotient (CQ), and the different contributions of "head, heart, and hands" project, program, and portfolio team members as discussed in Chapter 1, can assist in either amplifying opportunity or blocking threats. In the figure, the slices of Swiss cheese are meant to represent conveyors of threats. If the holes line up just right, they either allow or disallow the impending threat or the sparkling opportunity from traveling along to the project, program, or portfolio objectives and goals. Aside from the usual (and classic project control mechanisms, such as a risk management plan, use of solid risk identification practices, creating and using a risk register, represented by the three slices on the left), we have the ability to apply CQ (represented by the slices on the right) to block threats and amplify opportunities.

For example, if we think there may be a risk (remember that this could be either an opportunity or a threat) related to *alignment* with

---

* http://en.wikipedia.org/wiki/Swiss_cheese_model.

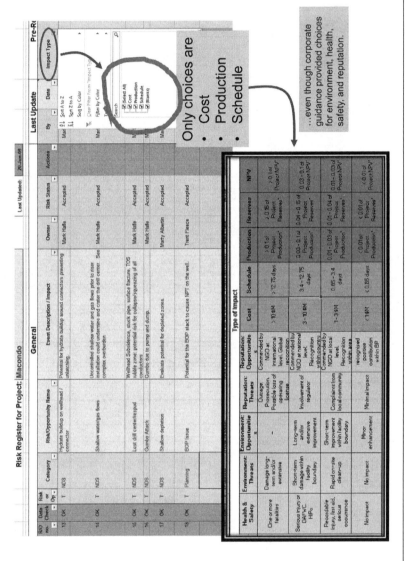

Figure 4.9   The risk register for the Macondo Well, with the template for risk identification from BP corporate.

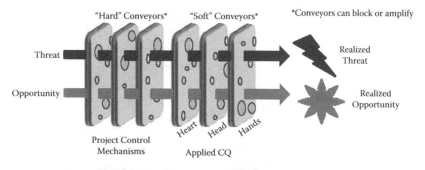

Figure 4.10 How threats and opportunities are amplified and/or blocked base on change style.

Desired project element	Change Style	Opportunity Amplification or Threat Blockage	Threat Amplification or Opportunity Blockage
Alignment	Heart	Acceptance	Dissension
Vision	Head	Understanding	Confusion
Control	Hands	Commitment	Rejection

a sustainability effort, we can count on our "Heart" people (coaches, champions, facilitators) to help convey acceptance, and we would want them to help squelch dissension with their particular capabilities. If the risk is regarding our *vision* for sustainability, we want our "Head" people (visionaries, drivers, and champions) to help either promote understanding of that vision or prevent confusion as it's deployed. And finally, if we are dealing with the familiar example of control—making sure that the rubber is really hitting the road—then, you guessed it—we need our "Hands" people (executers, drivers, and facilitators) to gain commitment to sustainability efforts and thwarting rejection of these efforts.

Now that we've given the background and context for detecting CSR threats, let's look at how your enterprise sizes up in this dimension.

Reject

Introduction

Waste! Reject it! This surrounding dimension of the sustainability wheel continues the process of being organizationally sustainable. This is important to the overall sustainability performance of an organization because it not only signals to (all) stakeholders that there is a *real* sustainability effort, but elimination of waste is a key to increasing the bottom line. Eliminating waste as well as wasteful behavior should be an organizational sustainability goal.

The main question being asked here is this:

Are there goals for eliminating wastes in your organization's EMS?

Nothing takes place in a vacuum. There is a significant interconnection between the many issues related to sustainability, for example, energy use can be connected to the emission of toxic chemicals, which are in turn connected to waste management, etc. The interactions between these forces can also become very complex, but considering the issues as a "system," at the very least, can trigger discussions about the interactivity of the sustainability forces resulting in creative methods for dealing with them as a whole. Considering the forces as a system also provides a high-level overview that should be kept in mind as you work through the solutions required to improve an organization's sustainability efforts. In addition, by applying the solutions, the result will be sustainability efforts to prevent, reduce, or at least address the negative impact that business can have on sustainability.

Case Study: Subaru of Indiana Automotive

In 2002, Subaru corporate challenged Subaru of Indiana Automotive (SIA) to achieve "zero landfill" by 2004. It seemed like a daunting task, eliminating all of the waste from a manufacturing facility, ever, let alone in 2 years. SIA took the challenge very serious because they wanted to be true environmental stewards, not only eliminating form the waste stream, but reusing what cannot be eliminated. By 2004, SIA was recycling 99.3% of its excess steel, plastic, wood, paper, glass, and other materials. The remaining 0.7% is shipped to Indianapolis and burned to help generate steam. In 2006, SIA recycled 11,411 tons of scrap steel, 1,537 tons of cardboard and paper, and 963 tons of wood—equivalent to conserving 31,040 mature trees, 31,572 cubic yards of landfill space, 711,631 gal of oil, and 10,759,000 gal of water.

The question is: how did they do it? It isn't an easy answer. In order to answer that question, SIA management turned to their employees and the question they asked was: what can we (SIA) do to eliminate the waste stream? In the first month of the program that question generated 268 ideas from employees as to how to do it. Every aspect of the manufacturing process was considered. One example of the employee input pointed to the manufacturing process. During manufacture, there was a small amount of excess steel that had to be trimmed. By recalibrating the machines, that excess was eliminated, saving 102 lb of steel per car, which translates to 425 coils of steel per year, or the equivalent energy to produce that steel of powering 2233 homes for a year.

The lubrication system for the engine parts required individuals using a spray bottle. By automating that process, in 2007, SIA eliminated 670,000 gal of oil. Sealing material is used to cover weld cracks during the painting process. That material was then cleaned off and lost down the drain. An additional step of scraping the material off and putting it back in the contained allowed continual reuse. Employees took it one step further and cut the container lid in half to use the edge to clean off their scrapers.

Going from incandescent lighting to compact fluorescent lamps (CFLs) causes its own issues. It is critical that the CFLs be disposed of safely. SIA has a novel solution to the problem they call the "bulb eater." The CFLs are "fed," excuse the pun, into the bulb eater and are reduced the various components, glass, metal ends, phosphorous, and even the small amount of mercury (that is particularly dangerous to the environment) so that they can be disposed of properly. By going to the CFLs, SIA saved enough energy to power 6000 homes for a year.

Recycling was another issue that needed to be addressed. Excess packaging around raw materials was a problem. One of the solutions was to ship all that packing material back to Japan for reuse. While expenses were incurred to ship the materials back, the expenses were recouped by not having to produce new packing materials as well as not having to pay for the cost of hauling the excess away.

This is a great example of approaching sustainability one issue at a time, yet tying it together with a system. This systematic approach is what drove SIA, and while leading to a "zero landfill," it became the first U.S. car manufacturer to be designated a Backyard Wildlife Habitat by the National Wildlife Federation.

There are a variety of ways that sustainability can be implemented. While organization may not be in the position to offset a new building against a positive effect elsewhere, but some creativity can lead to a very positive effect. As an example, some municipalities are willing to "trade" with developers, who are considering developments in vulnerable areas for town-owned land that may be less sensitive.

Some of the questions that can be asked of your organization are as follows:

- How much water is consumed?
- How much of it goes to wastewater treatment?
- If new facilities are being proposed, is the complete life cycle of their infrastructure being considered?
  - Is improvement of water consumption and treatment at design, operation, maintenance, and renewal of plant and equipment and buildings?
  - Are international standards that provide benchmarks as well as standards within industries and professions that affect individual industrial processes being considered?
  - Are buildings being designed or retrofitted with the U.S. Leadership in Energy and Environmental Design (LEED) certification being considered?
  - Is a BREEAM assessment for sustainable building design, construction, and use (developed by the Building Research Establishment [BRE] in the United Kingdom) being used?*

Whether a large or a small organization, top-down or grassroots, other questions can be asked in the following:

- Are measurements available for water usage? Establishes a baseline.
- Are low flow toilets/waterless urinals being used or considered?
- Are the fixtures being maintained?

---

* http://www.4-traders.com/SPARKASSEN-IMMOBILIEN-AG-6496284/news/ Sparkassen-Immobilien-AG-S-IMMO-AG-Green-building-certification-and-another-award-in-SEE-17000131/.

- Can we collect water and use from roofs or recycled wastewater for non-potable water use?
- Are employees being encouraged to report leaking fixtures or water usage abuse?
- Do we use compostable dishware and utensils in the cafeteria?
- Is our data center using closed system cooling?

Romilly Madew, chief executive, GBCA,* says the project's success was due to concentrating on nine areas:

- *Management*
  - The GreenHouse fit-out achieved Green Star points for the production of a tenant guide and waste management plan. Both garbage and recycling are weighed and recorded so that waste recycling percentages can be monitored. Currently, around 50% of all waste is being recycled.
- *Indoor Environment Quality*
  - An internal post-occupancy evaluation of the GreenHouse has found that 95% of all staff had a positive or very positive perception of their new workplace. As one staff member said: "The GreenHouse is by far the most professional indoor work environment I've experienced."
- *Energy*
  - The GreenHouse reinforces the claim that green buildings can routinely achieve energy savings of at least 20%–30% when compared with industry standards. The design intent of the GreenHouse was to keep energy use below 65,000 kW h per year. Lighting and power energy use are monitored and recorded on a monthly basis, and the results point to a predicted energy use of around 48,000 kW h per year, well below targets.
- *Transport*
  - The GBCA achieved the maximum five points for the GreenHouse's proximity to public transport. A further two points were allocated for limiting the number

---

* http://www.gbca.org.au/uploads/132/2436/Greenhouse%20case%20study_It%20is%20easy%20being%20green_270809.pdf.

of available parking spaces, further promoting the use of alternative modes of transport.

- *Water*
  - Water-efficient dual-flush cisterns, waterless urinals, and 6 Star water efficiency labeling standards (WELS)-rated bathroom taps have reduced water consumption. A 400 L gray water tank collects water from the kitchen taps, dishwasher, and hand wash basins, which is treated and then used to flush toilets—with 100% of gray water being reused on site.
- *Materials*
  - The workstations, walls and partitions, chairs, tables, storage units, and flooring used in the tenancy fit-out all achieved Green Star points for their reduced environmental impact. Where possible, ceiling installation was limited to avoid unnecessary use of materials. Instead, building services and cables are left exposed. Meeting room and work areas are flexible and adaptable spaces.
- *Land Use and Ecology*
  - The GreenHouse, which occupies space in an existing building, was awarded Green Star points for lease clauses committing to improve its environmental performance. The lease agreement includes quarterly energy, waste and water monitoring, and associated reduction targets. Cleaning products used to maintain the GreenHouse were chosen for their low environmental impact.
- *Emissions*
  - The GreenHouse's thermal insulation avoids the use of ozone-depleting substances in both its manufacture and composition.
- *Innovation*
  - The GreenHouse achieved an innovation point for the use of indoor plants—such as the installation of a green wall behind the reception area which provides both visual privacy and purifies the air. Additional innovation points were awarded for the gray water collection system, retrofitting a highly energy efficient displacement ventilation

system and closing the loop on organic waste composting within the office.

- The primary way to control waste and reduce the amount going to landfill or other methods of disposal is simply to produce little or no waste. This should be the objective of any business sustainability initiative. In Chapter 2, we saw how a carpet manufacture has been substantially reducing its waste and created technology to recycle old carpet from any manufacturer.

- Waste prevention is in fact the first of three principles adopted by the European Union in its approach to waste management:

  - *Waste prevention*: This is a key factor in any waste management strategy. If we can reduce the amount of waste generated in the first place and reduce its hazardousness by reducing the presence of dangerous substances in products, then disposing of it will automatically become simpler. Waste prevention is closely linked with improving manufacturing methods and influencing consumers to demand greener products and less packaging.

  - *Recycling and reuse*: If waste cannot be prevented, as many of the materials as possible should be recovered, preferably by recycling. The European Commission has defined several specific "waste streams" for priority attention, the aim being to reduce their overall environmental impact. This includes packaging waste, end-of-life vehicles, batteries, and electrical and electronic waste. EU directives now require member states to introduce legislation on waste collection, reuse, recycling and disposal of these waste streams. Several EU countries are already managing to recycle over 50% of packaging waste.

  - *Improving final disposal and monitoring*: Where possible, waste that cannot be recycled or reused should be safely incinerated, with landfill only used as a last resort. Both these methods need close monitoring because of their potential for causing severe environmental damage. The EU has recently approved a directive setting strict guideline for landfill management. It bans certain types of waste, such as used tires, and sets targets for reducing quantities of biodegradable rubbish. Another directive lays down tough limits on emission levels from incinerators. The Union also wants to reduce emissions of dioxins and acid gases such as nitrogen oxides (NO_x), sulfur dioxides (SO_2), and hydrogen chlorides (HCl), which can be harmful to human health.*

Organizations should think in terms of reduction, redesign, reuse, recycle, and composting as ways of reducing the amount of material entering the waste stream. Reduce, redesign and reuse can help, recycle, not so much. Recycling is the last "r" in the list and that is intentional. Recycling is an excellent method for keeping wastes out of landfills. However, it does not effectively reduce the waste stream, just gives the waste a different direction. Recycling requires an effort to pick up recyclables and process them. It is true that those materials being recycled can reduce the need for raw materials, but overall reduction of wastes is more effectively managed by redesigning and reusing products.

There are many ways of reducing an organization's waste. It may be as simple as mandating that printers must be set to print on both sides of a sheet of paper. Electronic communication can also reduce the need to print reports, time sheets, invoices, etc., while sharing copies, proof reading documents online, not printing drafts, and providing mobile workers with electronic notebooks/tablets can all help to reduce paper use. If every organization reduced the input of raw material, e.g., paper, it would have a much wider impact than reducing waste. Paper manufacturing is, for instance, "the largest industrial user of water per pound of finished product," according to the American Forest and Paper Association.

- Redesigning to reduce waste can take many forms ranging from redesigning administration forms and processes to use less paper or changing them to an electronic format. Manufacturers can redesign products with fewer parts, or invent new processes that generate less waste during the

* http://ec.europa.eu/environment/waste/index.htm.

production. Software simulations can reduce the number of prototypes or models that have to be produced.

- One simple way to reduce the waste stream is to remove disposable coffee and water cups and replace them with reusable ones. There might be a slight increase in water usage but the benefits can outweigh the costs.

- Recycling, while not the most effective method for managing the waste stream, is none-the-less, a great way to reduce the waste going to landfill. Nexan is a world leader in the cable industry. At any one time, the company has more than 350,000 wooden cable drums in circulation within the European Union. To advance their commitment to sustainability, Nexan has become the first cable manufacturer to upgrade to a "green drum." The old drums are designed to be reused a number of times and according to their age and condition, be replaced each year. The new drums, also being reusable, will be certified by PEFC™ (Programme for the Endorsement of Forest Certification). The program started in 2010 and continues today. That is one example of how an organization can combine recycling and a more sustainable product.

- Further, an organization's procurement process should assess supplies against criteria that include whether they can be recycled. The list of potential consumables that can be recycled continues to expand. They include printer cartridges, non-potable water, cleaning solutions used in manufacturing, computer and office equipment (repurposed, refurbished, and recycled), glass and metals, building materials during renovations, and more. Another simple action is to provide desktop recycling containers for employees, placed near printers/photocopiers, etc., and to purchase paper that can easily be recycled, i.e., without adhesive, bright colors, or bleaching.

- Composting, an area more normally associated with consumers, is another way to reduce the waste stream from an organization. If there is a cafeteria, then leftover food products (except meats, fish, fats, and oily food scraps) and appropriate dishware and utensils should be composted. The organization can either compost its own waste or have it taken away.

- Another way of reducing an organization's waste stream is to ensure that vendors either reduce packaging materials, consider multipacks, or take packaging material back. If you are also a supplier to others, then this of course applies in reverse.
- The ability to control an organization's waste stream efficiently has far-reaching, positive consequences across a variety of environmental concerns. Controlling your organization's waste stream is not only the right thing to do, it is also doing things right.

In Chapter 2 of *Green Project Management*, we detailed the four principles of sustainability of The Natural Step™. Two of them are relevant here as given as follows:

- "Eliminate our contribution to the progressive buildup of substances extracted from the Earth's crust (for example, heavy metals and fossil fuels)"
- "Eliminate our contribution to the progressive buildup of chemicals and compounds produced by society (for example, dioxins, PCBs, and DDT)"*

It is the goal of business sustainability to reduce the amount of raw materials extracted and to eliminate the production and waste of destructive products, or by-products, from those materials that are extracted.

From a global perspective, the production of greenhouse gases (GHGs) can be considered waste and is an area of concern, and a reduction in energy use will not only save money, but also help reduce the growth in emissions. Many industrial processes use hazardous substances and produce hazardous waste. This has led to legislation and regulation that has eliminated the use of some materials (asbestos, for example). There are areas of the world where legislation is less effective and some companies have used loopholes to use processes that would not be tolerated elsewhere, or claim a lack of knowledge about conditions at the end of long and complex supply chains.

_____

* http://thenaturalstep.org.

Responsible organizations have been tackling the way in which they design or manufacture products.

Case Study: AT&T

In the late 1980s and early 1990s, AT&T was using a particularly caustic substance called perchloroethylene (PCE), which was used to remove flux after soldering.

When faced with the increasing environmental damage caused by the use of this chemical as a by-product waste and a desire to become more sustainable, Bell Laboratories' scientists "found that a low-solids flux, when applied in even and controlled amounts, would not leave the tacky residue that required cleaning. Rather than substituting another solvent for PCE, the engineers redesigned the soldering process so that cleaning was no longer required."*

The other interesting thing about this study was that it "eliminated the 35,000 gallons of PCE it purchased each year and also eliminated the need for related industrial hygiene and environmental monitoring activities, and generated cost savings of $210,000."

Another way to reduce or eliminate the use of toxic materials, thus toxic wastes, is to look at "greening" the supply chain. This entails having answers to a number of questions from your suppliers and ensuring they do the same for their suppliers. You need to know:

- What are *their* environmental values?
- How do they measure and enforce those values?
- Do they have an environmental management plan (EMP)?
- Do they have a chain of accountability for the EMP?
- Do they understand your environmental values?

Another example is the use of carbon trading and offsetting (as we talked about in our previous book). It does have arguments both for and against: it is often criticized for exporting a problem rather than solving it. A similar concept for biodiversity could be subject to the same criticisms but organizations could adopt a strategy of contributing either monetarily or in resources to organizations like the Nature Conservancy in the United States, or similar organizations worldwide.

---

* http://pdf.wri.org/bell/case_1–56973–125-X_full_version_english.pdf, p. 8.

Support of school programs like Science Technology Engineering and Mathematics (STEM), Backyard Wildlife Habitat, like SIA did, and conservation groups or others involved in preserving natural habitats influence an organization's sustainability.

So whether you are actively pursuing your *rejection* strategy; reducing or eliminating your waste stream, implementing energy reduction, lessening your environmental impacting GHGs, or buying carbon offsets to help others defray the costs of their rejection strategies while your organization continues to pursue rejection strategies, the reject surrounding dimension is critical in completing the cycle of sustainability.

Project

Introduction

Project (prō-ject) is that surrounding dimension that contains the opportunities available to the "sustainable" organization and opportunities available to organizations where sustainability is a work in progress. The main question here to ask is as follows:

Are we aware of the opportunities created by being more sustainable?

Maturity Models

Project management maturity models are ways to assess the maturity of the organization's project management functions. One of those models is from the Project Management Institute (PMI). "The Organizational Project Management Maturity Model (OPM3®) is a global best practice standard to assess and develop capability in Portfolio Management." From pmi.org, "*OPM3* offers the key to organizational project management maturity with three interlocking elements:

- KNOWLEDGE
- ASSESSMENT
- IMPROVEMENT

Those elements are defined by PMI as researching and understanding industry best practices (knowledge), evaluating an organization's current situation and identifying areas of improvement (assessment) and next steps to continual improvement (improvement).

Another interesting maturity model, the French Software Engineering Institute (SEI) in collaboration with MITRE Corporation, has refined the Capability Maturity Model (CMM). This model can easily be applied to project management. For the purposes of this document, we have modified the process to be more relevant to the sustainability maturity of an organization.

Figure 4.11 shows an example of a "key practice" to help the organization in their planning to allow them to take advantage of sustainability opportunities. Maturity models are an important tool for the sustainability management of an organization because using those models will help assess where the organization is relative to its maturity and, more importantly, where it needs to go.

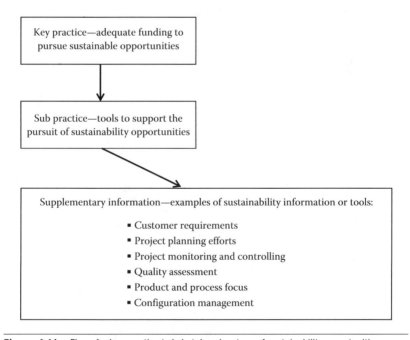

Figure 4.11   Flow of a key practice to help take advantage of sustainability opportunities.

Opportunities and Challenges

Business Cases

According to a recent study, *The Business of Sustainability*, by MIT Sloan Management Review and The Boston Consulting Group (sloanreview.mit.edu/reports/the-business-of-sustainability/) "the biggest drivers of corporate sustainability investments are government legislation, consumer concerns and employee interest in sustainability." The study also finds a contradiction "that sustainability professionals find quantification of the business case difficult." As mentioned earlier in the maturity model section, organizations reach a higher "sustainability" maturity when they understand the opportunities. The corollary to this is that higher maturity organizations are more likely to be able to take advantage of opportunities because they are not concentrating on gaining maturity. Less mature organizations do not have a structure in place to pursue opportunities like reducing GHG emission, eliminating their use of toxic chemicals, reducing packaging, or even recycling and reuse. A primary reason for this is the lack of understanding as to how this can positively affect the bottom line, or in other words, the question: What is the business case for sustainability?

Environmental and Sustainability Education

Environmental and sustainability education (ESE) can be both an opportunity and a challenge to an organization. The challenge to education is the ability to measure its impact on an organization. It is more of a "soft" measure connected to the opportunity. ESE can be a great motivator to employees interested in becoming more sustainable. That interest is sparked by the organization's top-down commitment as well as some innate quality of the employee. Coupled with engagement, ESE can be a great motivator for employees because it can not only be applied to saving resources ($) at work, but can also be translated to everyday life, home life. According to *www.ecomii.com/business/green-workforce*

> Human capital (a company's employees) and the systems surrounding it
> are the true cornerstones of building a sustainable business. Sure, you

can argue that businesses are sustainable because of their operations and culture, but it's the employees who craft and execute those eco-policies and create that green corporate culture. Without developing personnel and implementing sustainable strategies, it's rather difficult to go green successfully. That's why human resource (HR) practices are a key component of sustainable business development.

Further, ecomii.com gives four reasons for having sustainable-oriented employees because of the following opportunities:

- *More inspired problem solving:* Employees who bring a sustainability lens to business decisions allow for a broader perspective that sparks innovative solutions to both common and newly emerging climate change-related business problems.
- *Increased desirability as an employer:* Intellectually knowing what sustainability is and practicing it in daily decision making are two different animals. As you become known as a desirable green employer, you'll have your pick of the green talent pool—individuals who already understand sustainability and have practice in maximizing people, planet, and profit through business strategy. Bringing them onboard gives you a powerful market edge. Just look at Patagonia, a company that receives *thousands* of applicants for each posted job opening. The synergy that builds from green-minded employees working together in a business can be unbelievable.
- *Less stressed budget:* Many employees who are committed to sustainable careers are amenable to flexible compensation and benefits, often preferring alternative transportation, flex work schedules, and other low-cost benefits over hard dollar cost-of-living increases. These options can give you more bend in your budget.
- *Improved employee retention:* Many green companies these days boast low turnover rates compared to their non-sustainable counterparts. That's not just talk. In a green workplace survey conducted by the Society for Human Resource Management (SHRM), 61% of respondents who worked for an environmentally conscientious company said they were "likely" or "very likely" to stay at the business because of those practices.

Some of the questions that could be asked in support of ESE are as follows: Do we have an ESE infrastructure? Are we taking advantage of our education infrastructure to encourage more sustainability?

Profit

We've been following MIT Sloan Management Review and The Boston Consulting Group's sustainability reports since beginning our research for our previous book, *Green Project Management*, CRC Press, 2010. There are some very interesting observations from their 2013 report, "Sustainability's Next Frontier, Walking the talk on the sustainability issues that matter most." Organizational capabilities play an important role in the ability of organizations to take advantage of opportunities. The study found that two-third of the companies reporting significant success in addressing sustainability issues (or taking advantage of the opportunities) have strong support from leadership, where only one-third of the companies having "somewhat" or "barley" addressing sustainability issues have the same level of support.

Another point made in the study is about profit, one of the three ps. In the report, nearly 60% of those companies addressing sustainability issues saw increases to their bottom lines, where only 19% of the companies "somewhat" or "barely" addressing issues showed any increase. Within the MIT/BCG report is a 2013 Harvard Business School report, by professors Robert Eccles and George Serafeim, analyzing financial performance of "high-sustainability" companies. The analysis shows that if you invested $1 in a portfolio of high-sustainability companies in 1993, that by 2010 that investment would have grown to $22.60, while if you did the same in less sustainable companies, the investment would have grown only $15.40, a significantly less return on investment.

Partnering

One of the more interesting and nonintuitive opportunities for organizations is their partnering with other like-minded organizations, even if they are competitors. The 2013 MIT/BCG report uses the example of Nestlé. They have "turned to customers, advisors, and competitors to develop what it calls 'precompetitive' practices." The opportunity was to use this type of collaboration to address common issues like child labor and pesticide contamination. Because of the size of these issues, it didn't make sense to try to address them alone, but rather to, according to Hans Joehr, corporate head of agriculture, to reach out to Danone and Unilever to work together and "develop (common) principles and practices and procedures."

McKinsey.com (/www.mckinsey.com/insights/sustainability/creating_partnerships_for_sustainability) provides some additional insights into the opportunity to partner with other organizations. "The effort needs to help each partner organization achieve something significant. Incentives such as 'we'll do this for good publicity' or 'we don't want to be left out' are not sufficient."—*Nigel Twose, Director of the Development Impact Department, International Finance Corporation, World Bank Group.* It's more than just saying yes because you don't want to say no, there have to be other motivational reasons, like people, planet, profits (three ps). According to the article, whether the motivation is one of the three ps, "enlightened self-interest is the only sustainable motive."

Large issues may be easier to address than the smaller issues. For instance, the article points out that the collapse of the North Atlantic's Grand Banks fishery in the 1990s "made the fisheries industry more interested in sustainable harvesting practices."

Partnering is not easy, especially when the partnership may involve a competitor. "For the greater good" may not be the best answer. It is important to look for concrete reasons, particularly around enlightened self-interest.

Are we partnering with other companies, especially for the larger sustainability issues?

Strategic Value Creation

Table 4.4 captures the opportunities and loss of opportunities when strategic value creation is not pursued by the enterprise. There are several points in Table 4.4 that we found particularly interesting. The sources of value creation are important concepts. They reemphasize the points made earlier; attention to issues, attention to enterprise reputation, alignment with stakeholder value, ensuring that CSR

Table 4.4 Benefits and Costs of Strategic Value Creation

SOURCES OF VALUE CREATION	BENEFIT/OPPORTUNITY OF ENGAGING	COST/RISK OF UNDER-ENGAGING	WHAT'S POSSIBLE
Resilience: Tracking sociopolitical and environmental issues.	• Issue Identification • Preparation: Mitigation and adoption • Co-creation and collaboration on solutions	• Absence or loss of trust • Lack of preparation for crisis • Negative media • Costly cleanup • Damage control • Stock market losses	Resilience: the ability for both the business and its operating environment to resist impacts
Reputation: Monitoring and managing stakeholder expectations	• Reputation capital • Trust • Network of third-party reputation defenders	• Absence or loss of trust • Unmet expectations • Crisis and damage control • Negative media • Stock market losses	Renewable Reputation: a potentially limitless source of reputation capital
Alignment: Understanding stakeholder values and ensuring CSR program impact	• Optimize and validate program investment • More effective and measurable impact • Increased budget • Reinforcement of results	• Ineffective impact/results • Underperforming financial investment • Demotivated team • Difficulty justifying budget	Virtuous Value Creation: alignment and stakeholder relevance increase measurable results, investment, and in turn, social good
Strategy: Sourcing the wisdom of the crowd and co-creating solutions	• Innovation • Differentiation • Capture market opportunities as they emerge • Co-create and collaborate	• Missed business opportunities • Loss of market share • Stagnant revenue growth • Un-utilized source of thought capital and initiative	Sustainable Competitive Advantage: a generous source of ideas to improve business outcomes

Source: www.greenbiz.com/blog/2014/10/23/stakeholder-engagement-key-csr-online-communities. Used with permission.

programs are visible and impactful and that the strategy takes advantage of collaboration and sharing.

The benefits and opportunities to the enterprise are significant. Issue identification is critical to the enterprise's overall understanding of the business environment in which they operate. The concept of reputation capital is important in that it can "insulate" an enterprise from a mistake, giving them enough time to react and make a course correction. A continual emphasis on gaining reputation capital can ensure that there is enough focus on reputation capital – and propensity to act on it when it is needed. We're not saying that this is the only reason to pursue reputation capital. We are saying that there can be mistakes made in a sustainability effort. When an enterprise is working as hard as it can to be more sustainable, "banking" reputation capital can provide that cushion, or at least help mitigate the problem. And, with "a network of third-party reputation defenders," it will be easier to get back on track. The defenders can provide additional protection, again, to allow time to recover.

Building stakeholder trust leads to many benefits: increased market share, increased budgets for sustainability efforts because of the increased value due to increased market share, innovation to continually improve sustainability efforts, and increased opportunities to work with other like-minded organizations. Making a point of sharing their sustainability efforts, Gary Hirshberg, *Stirring It Up; How to Make Money and Save the World*, Hyperion, New York, 2008, Yvon Chouinard, Patagonia, Ray Anderson, Interface, and Jeffrey Hollender, Seventh Generation, collaborated with MagicGreen Productions on an award winning documentary, *so right so smart* available online at sorightsosmartfilm.com.

It may be costly to not create value through sustainability efforts. Most of the good things that have to do with creating value could be lost. Greenwashing (lack of trust) can occur. There may be violation of regulations resulting in penalties: fines, loss of reputation, unmet stakeholder expectations including loss of market share and demotivation of employees, loss of budget because sustainability is perceived to have no value, and missed business opportunities.

With value creation, the possibilities can be limitless. Sustainability issue impacts can be minimized, reputation can grow, stakeholders (customers and employees) seek out the enterprise, and business outcomes improved. For some organizations, sustainability value creation is not a priority. We believe that type of thinking is short term. Perhaps in the short run, there will be additional capital created by not pursuing sustainability value creation. In the long run, however, we believe that it is a mistake and "un-sustainable," excuse the pun, not to pursue the opportunities afforded by sustainability value creation.

Case Study: Walkers Crisps

Walkers Crisps is a British snack food similar to American potato chips. It is a subsidiary of PepsiCo. Walkers did a study to see how they could be more efficient in producing chips. In effect, it was an Life Cycle Assessment (LCA). They were paying farmers by the weight of the potatoes they received. So naturally, the farmers did whatever they could to make the potatoes heavier. They kept them wet, using humidity chambers that required power and supervision but it paid off in terms of the weight it added. Then the potatoes were shipped into the manufacturing facilities and the need for larger trucks to carry the heavier loads. The larger trucks required more fuel to power them. Once the potatoes were in the manufacturing process, the extra moisture had to be burned off during the frying process, which took more energy.

The conclusion was that simply by changing the measure to dry weight, and paying the farmers accordingly, the farmers did not need the humidity chambers to keep the product wet, saving money on their energy costs. The company was now paying for shipping on less weight, and could use smaller, more fuel efficient trucks, saving money on fuel. Once the potatoes were in the manufacturing process, less energy was needed to fry out the extra water. It was a win–win opportunity for the farmer as well as the enterprise, in terms of both financial and environmental impact.

Case Study: Shell Oil

In our presentations on Sustainable Project Management, we use this quote: "When I look at an investment proposal now," says Marvin Odum, president of Shell Oil Co., "it still covers the technical issues, of course. It certainly covers the financial issues. But fully half of that proposal deals with what I would call the nontechnical risk: *social performance* and *sustainability issues* (authors' emphasis)."

"Here's where it gets even more interesting," he continues. "As you get better companywide at exploring, understanding and addressing those nontechnical risks, it drives innovation. Because mitigating those risks often drives you right back into the technology loop—back into asking how can you solve novel problems in novel ways, and how can you do it at affordable cost? At this point, it may be the number one driver of our innovation program."

According to *MIT Sloan Management Review*, Spring 2011, Like other energy companies, Shell is in a classic "between a rock and a hard place" situation. The world wants what Shell provides, but it wants it when it wants it, at a price it likes to pay, and with positive or at least neutral environmental and social impact. That's forced the company to adapt its traditional innovation approach—and even its overall organizational structure—in some surprising ways.

The need for those changes has also been heightened by the environmental damage and public relations disaster of the BP oil spill in 2010. Odum says, "What the Gulf of Mexico spill shows us is we are dependent on how the whole *industry* performs; it affects a part of our license to operate." That is true even though Shell enjoys a reputation for sustainability performance that is stronger than that of most other energy companies. For the last 2 years, respondents to *MIT Sloan Management Review*'s annual sustainability survey have named Shell Oil among the top 10 companies that are "world class" at paying attention to the issue. Still, dealing with the broader public perception and wariness that greets energy companies, says Odum, has become a major focus of the company. Today, managing the concerns of external stakeholders has prompted changes in management approaches and strategy internally, and sustainability issues have moved in Shell from being a company "priority" to a "core value." This is clearly an understanding of the opportunities and using them to become a better company for it.

Case Study: Marks & Spencer–Beyond Plan A

Recognizing that consumption is increasing worldwide in leaps and bounds and also realizing that the way business is conducted are unsustainable (to quote their website corporate.marksandspencer.com/blog/stories/why-marks-and-spencer-is-proud-to-join-collectively-a-new-global-sustainability-movement, "A growing population and burgeoning consumption is an exciting growth opportunity for the consumer goods market but there literally are 'not enough fish in the sea' (or wood in the forests, soil on the land or water in the river) for everyone to live as we do in the developed world."), Marks and Spencer, a major British multinational retailer, is partnering with Unilever, BT, a British communications company, Coca Cola, Carlsberg, Google, Facebook, Twitter, Audi, and Pepsi focusing on consumers and "Collectively will showcase all that's great about the future showing we can have a good lifestyle that's in harmony with the needs of nature and communities." The collaboration is known as *collectively*.

In 2007, Marks and Spencer launched a program called Plan A ("because there is no Plan B"). Plan A (2007) consisted of 100 sustainability commitments to be achieved within 5 years. They have now launched Plan A 2020 with 100 additional commitments. Again from their website, "Through Plan A we are working with our suppliers and employees to inspire our customers, be in touch with the communities we depend on to succeed, innovate to improve things for the better and act with integrity."

Marks and Spencer are an example of a company that not only look internally for their sustainability value creation and opportunities, but also use partnering and collaboration to seek out external opportunities.

5

THE ROAD

Introduction

Chapter 5 is a chapter that begins to bring the Sustainability Wheel™ from theory to practice by engaging the "outside" world with the various "languages" of your enterprise. It is the "road" in "where the rubber meets the *road*." It contains three surrounding dimensions: dialect, intellect, and circumspect. The dialect dimension is getting the word out beyond your enterprise. The intellect dimension is looking outside of the enterprise and benchmarking against other organizations looking to achieve a more sustainable organization. It also includes sharing information, understanding external forces (regulatory as an example), and, in general, a better understanding of the "triple bottom line." The circumspect dimension is the "look back" to ensure the connection with the respect dimension.

Dialect

"Not only the entire ability to think rests on language ... but language is the crux of the misunderstanding of reason itself" is a quote attributed to Joann Georg Hamann, a German Philosopher from the 1700s. While it may be a slight exaggeration, without accurate communications in a "language" we understand, the "message" is lost. The dialect dimension tests the hypothesis that the organization is being understood by the external environment (external to the enterprise) and acknowledged for its efforts. As attendees at the Project Management Office (PMO) Symposium in 2013 in San Diego, we were intrigued by comments from Project Management Institute (PMI) Chief Executive Officer (CEO) Mark Langley, in which he said that in his background as a member of the boards of directors of several large corporations, he never heard project management terms—Gantt Chart, Work Breakdown Structure (WBS), and so forth—during board meetings. In fact, he rarely, if ever, heard the word "project" come up. The focus was on the operations of the companies—the deliverables in their steady state. Of course, we all know that it's the project and program managers, who enable these deliverables, but in order to communicate between the PM community and executives, it's important to have a common language—and it's not PM language. At that level—where the enterprise is really run—the language, of necessity, is business language.

This conference also reinforced our increasing view that the project level and practitioners of project management were not necessarily the correct audience, or at least not the *only* audience for the message about sustainable thinking. So, many of the themes at the PMO Symposium were about strategy, mission, vision, and values, themes that were nearly absent at the PM conferences we've attended.

The main question to ask here is as follows:

Do we understand the external sustainability environment so that we can "speak" the language?

Additionally, is the organization open to sharing its efforts with others? Does the organization understand the external environment (regulatory as an example)? Are we doing whatever is necessary to understand the triple bottom line, the people, planet, and profits?

Communications

Are we communicating our sustainability efforts outside our enterprise? If we are communicating our sustainability efforts are they being understood and acknowledged? The first question is probably easy to answer; the second question will take more effort.

The ability to communicate is one of those "tools" in the project manager's toolbox. However, what we sometimes don't realize is that not everyone has that tool. Even if they do, sometimes it is using a hammer when a screwdriver is needed. What you are really trying to do is connect with the outside world. Because your enterprise is "on board" with your sustainability mission, the object in the dialect dimension is to convey that same sense of unity with your mission.

Tools

We are firmly ensconced in the "communications age." There is nowhere to get away from it except in the remotest parts of the world, and probably not even there. Communicating is more than Twitter, Facebook, websites, emails, and more; it is about understanding the emotions of the receiver and sender. It is especially true in communicating sustainability efforts to all stakeholders*—who may have a particular emotional attachment to the message. Communications is all about the sender (person or enterprise) and the receiver of that information. From an idea, the sender formulates a message and transmits it over a particular channel. A channel is too broad to define here, but suffice to say it can be any form or wired or wireless communications. The problem is that each and every channel may have its own emotional issues attached to it. Once the message is transmitted across the channel, the receiver then listens to it and extracts some meaning based on the listeners' abilities. Those abilities are utilized to formulate a response, which is then sent over the channel to the sender via a feedback loop.

Let's talk emotions. Take a Technology, Entertainment, Design (TED) talk for instance. One of the best TED talks about sustainability was back in 2009 when Interface Global leader Ray Anderson[†] TED talk centered on Interface Global's climb of "mount sustainability" and their efforts to undertake "Mission Zero." Without the emotions shown by Mr. Anderson, the talk would have not been as effective. Emotion can be conveyed though the message, also. Words that denote passion—words that can paint a picture—are highly effective in communicating emotions. Visual people are good at conveying their passion, because they can "paint the picture" in words. They can "see" what they want to say.

Websites traditionally include pictures in an effort to convey part of the message. Isn't that a way of emotion? As an example, the American Society for the Prevention of Cruelty to Animals (ASPCA)

---

* I use the word stakeholder as anyone remotely connected to the communications of the enterprise from the CEO to the casual external reader.
† Ray Anderson has since passed away, but his legacy lives on.

runs television spots to raise money. Those TV spots are accompanied by pictures of dogs (particularly) that show the dogs in sad situations, tied to dog houses in bitter cold, etc. Those pictures not only "tug at our emotional strings," but also convey the emotions of the ASPCA. We're not saying that every website includes pictures the devastation in places like Hua, China, where the disassembling of computer circuit packs has rendered the area's living conditions deplorable, but choice of pictures is critical. Pictures will capture emotions.

Emotions are a very important part of our communication process. It is the feeling aspect as opposed to the thinking aspect. It can be the most difficult aspect of our process. It can be along the following spectrum (Figure 5.1):

At the "knee-jerk" end of the spectrum there are frustration, misunderstandings, and conflict. At the controlled end of the spectrum are clarity, understanding, and communications. While your message should be emotionally driven, being on the knee-jerk end of the message can obscure the intent, the way people understand it. Where is your enterprise along this spectrum with respect to communicating your sustainability message?

Additionally, some senders are not even aware of their own emotions. Acting in a certain way is just second nature. Within the nurturing environment of an enterprise, that may be fine, but when trying to communicate such an important message as sustainability efforts, not being aware of your own emotions is not fine. Two things: the sustainability effort itself is time consuming and probably costly to the enterprise, and therefore, the message is very important. Being aware of your emotions is not only good for you, but also provides insight to the emotions outside your enterprise making it easier to connect. If you stay strictly to the rational aspect of your message, you may fail to fully understand what others are hearing and miss the opportunity.

Knee jerk   Controlled

Figure 5.1   The spectrum of reactions to a message.

The following are some steps to help improve communications by being emotionally aware:

- Do some research to assess the external sustainability "climate"
- Understand your enterprise's sustainability climate
- Stay positive on your message
- Communicate clearly and effectively

"Once something is a passion, the motivation is there." Michael Schumacher, German racecar driver, retired. Another way to communicate with emotions is to communicate with passion: expressive, focused, and motivated. Where does passion come from? It is primarily from a desire to make something happen. To be passionate about your enterprise's sustainability efforts, all you need is to believe in it and the desire to "tell the world." Sounds easy, but with everything else that is going on with your work life, it may be difficult to generate that passion, especially when those efforts negatively affect the bottom line. Remember, however, that the hit to the bottom line is probably just short term and the long-term results will likely be significantly positive. Also, there is more at stake than just the profit bottom line. Profit is only one "p."

Communicating with emotions is only one aspect of getting your message out effectively. There is also the intellectual or thinking piece of the message. While thinking with the added benefit of emotions may become second nature, it will take some time. If you are not there yet, then you also need to be aware of the separation between emotions and intellect.

Effective communications require both thinking and feeling, and to paraphrase the Serenity Prayer, the wisdom to understand the difference. When your emotional response is solidified then you can have a good balance between your intellect and your emotions, which is your ultimate goal of effective communications. Are we conveying our message with emotion and intellect?

Another aspect of whether or not your "language" is being understood is whether the message communicated by the receiver is understandable. It is a test to see whether or not your "language," as the sender, was understood by the receiver. So beside the "speaking" skills, you also need listening skills to answer the question, are we speaking the same language? Do I possess the appropriate skills to really listen?

Listening is another communications skill. It is one of the more important aspects of effective communications. It understands not only the words from the receiver but also the emotions (there's that word again) that are part of the message. As pointed out earlier, emotions are not only conveyed verbally, but passion comes through the written words also.

Listening is "easier" when there is a face-to-face contact. By easier we mean that the ability to detect the nonverbal clues to the passion of the sender is more readily obtained. When one can focus on the speaker, many times the passion of the speaker is, therefore, apparent through their body language. However, to be a good listener is to avoid distractions, sometimes difficult in this communication age (text messaging, email, etc.). One method to force concentration is to repeat in your head, their words. Note-taking may also help, but it could also be a distraction. One of the things that happen, however, is our tendency to formulate the response before we've fully heard the sender (before the sender is through sending, perhaps). That can lead to interruptions because we want to get "our" word out. Another major distraction is judging both the sender and the sender's message, again, before the sender is finished or before we've given ourselves enough time to process the message. And, the receiver also has body language that may need controlling. Encouraging senders with your body language helps the sender relax and keeps you "in the game." Passion is about the degree to which we care. Senders with passion indicate that the subject is important to them.

But how do we "read" emails, texts, memos, reports, etc., for passion. Probably the most obvious is word choice. Look for active verbs: confront, demonstrate, exhibit, impose, focus, predict, and solve, as examples. Additionally, descriptive adjectives: abundant, disdainful, insightful, malevolent, oppressive, and perceptive can indicate passion, both positive and negative. Passively written verbs, particularly with past tense, indicate a less passionate approach to the issue. According to USC Aiken Writing Room (www.usca.edu/asc/pdf/writing%20 room/actpasverbs.pdf), "When a verb is in the active voice its subject acts or does something. When a verb is in passive voice its subject receives the action or is acted upon." When we are communicating are we doing so actively, with passion? Are we listening? Are we listening for the passion of the sender?

Dialect is about the language and the ability of the sender and receiver. Depending on the circumstances, we can be both! The different subcultures within an enterprise will require different dialogues. The focus for strategic managers will be more visionary. Their concern will be with innovation, differentiation, and market share. They will be listening for ways to use sustainability as a differentiator from competitors, looking for gaining market share by using sustainability efforts as differentiation, new technologies that will help the effort, and long-term bottom-line savings. Operations managers will be listening for more short-term gains, and how sustainability will make their jobs easier, or at least more efficient. The important thing to remember is to understand the audience and "speak the language" for maximum gain.

Intellect

The intellect surrounding dimension concerns the enterprise's efforts to benchmark their sustainability efforts and to strive for continuous improvement.

Are we benchmarking our sustainability efforts?

According to http://www.referenceforbusiness.com/management/A-Bud/Benchmarking.html:

"Benchmarking is the process through which a company measures its products, services, and practices against its toughest competitors, or those companies recognized as leaders in its industry." To paraphrase, benchmarking is a tool to help determine whether the company is performing particular functions and activities sustainably (or with sustainability), whether the costs of being more sustainable are in line with competitors doing the same, and if they are considering the internal activities and business processes that may need improvement. That improvement will be an effort to be more sustainable.

When using benchmarking for sustainability efforts, to use The Natural Step's™ four conditions, the enterprise can focus on its raw materials and suppliers, its outputs, its overall environmental impact, and its impact on people. It can focus on the roles of people within the enterprise and their job functions, or focus on its long-term planning. It all helps to establish the enterprise's sustainability mission and identify any processes and procedures that may be impeding its progress. Benchmarking can also help determine the "boundaries" of sustainability mission. On one extreme, will the sustainability mission be all encompassing, or will it address a specific issue like suppliers (upstream) or customers (downstream)?

Benchmarking is a powerful tool for assessing the sustainability within an enterprise. When used, it compels the enterprise to measure its own sustainability efforts against an external standard. It is a way to ascertain which competitors are the most "sustainable" or have the best sustainability practices that can be imitated. Why reinvent the wheel if there is something out there that will work? Not only can we determine what their best practices are, but we can use those practices and improve on them for our own efforts. For example, the research may reveal how an enterprise is "greening" up their supply chain. We may be able to use that information not only to green our supply chain, but also see whether or not there are alternatives that are more cost effective. We may also be able to better structure our employee sustainability training to be more effective by comparing it to companies who are particularly successful in their employee sustainability education and training. In every aspect, an enterprise's sustainability efforts may be improved by benchmarking.

It is clear that all efforts to improve an enterprise's sustainability can be undertaken all at once, and benchmarking can provide information about where the "biggest bang for the buck is" relative to opportunities for improvement. It can also shorten the implementation cycle of sustainability efforts by identifying the areas of concentration. When looking at the various tools available to the project manager, one quickly comes to mind to aid in this process. The work breakdown structure is one of those tools that can be used by the project manager to identify the tasks.

Are We Doing Research on Companies Similar to Ours for Their Sustainability?

The purpose of benchmarking is not only to look at similar companies to determine best practices, but also to look internally to access enterprise sustainability weakness that can benefit from the benchmarking. The objective is to elevate those weaker functions to the ultimate goal of "better that existing best." For example, it is the ability to use the strengths/weaknesses/opportunity/threats (SWOT) model to identify the weaknesses in sustainability efforts and move them to strengths and to identify the threats to the sustainability of the enterprise and turn them into opportunities.

Are we using a SWOT analysis as a tool to consider our sustainability efforts, to find innovation, continuous improvement, to find areas of employee engagement improvement, and competitive advantage?

One of the major areas for improvement for project success, for example, is the ability to clearly define roles and responsibilities. Benchmarking can help the enterprise focus on the various roles and responsibilities designated by best practices.

Are We Using Benchmarking for the Right Reasons?

To answer that question, you need to know why you are doing the benchmarking and the answer to the "why" must be appropriate to your sustainability efforts. Some of the appropriate reasons to pursue benchmarking as part of the enterprise's overall sustainability strategy are to indicate to employees that management is willing to undertake a proactive approach to sustainability. It established the important goals and objectives that are so critical in understanding the sustainability mission. Not only is establishing those goals and objectives success factors, but also being able to focus on clear goals and objectives enables the enterprise to seek those potentially high-payoff opportunities that may have been otherwise missed, as mentioned previously. Another "why" is to highlight the fact that the sustainability effort is based on the facts gathered from other similar efforts. Altruism is a great reason for a sustainability effort, but concrete reasoning is even better.

Using different sources of information, speaking with the experts in your industry, consultants, customers, and employees can increase the effectiveness of benchmarking. The Internet is a great source of information. Companies within your industry have extensive information on the websites. Quantitative and qualitative analysis of industry data will also provide a benchmarking baseline.

To be successful in using benchmarking for your sustainability efforts, it is the *usual suspects*. Management has to be fully engaged not only in the process, but in the implementation. What good is collecting all of the information if it stagnates because of lack of management commitment? Commitment of management includes having the time and resources to implement the sustainability best practices. *Do we have management commitment?*

Are We Sufficiently Educated in the Benchmarking Process?

As we've said earlier, benchmarking is a very powerful business tool. It can and should be applied to sustainability. It can support the business case for sustainability. Benchmarking can improve the effectiveness of an enterprise's sustainability efforts. It can help gain stakeholder commitment and satisfaction. Benchmarking should be a part of the enterprise's overall sustainability effort.

Circumspect

In Chapter 1, we referenced the plan, do, check, act cycle, and Dr. Cynthia Scott's sense, scout, synthesize, steer adaptation of that model. Here, we imagine that model as a spiral rather than a cycle (see Figure 5.2). The very fact that you are reading this book puts you on the right path to follow this spiral of success. Each time you go through the cycle, the enterprise gains more knowledge, more experience, more connectedness, and more wisdom.

With respect to the assessment here in the book—and the ones to which we refer in the connect chapter, the key is to look for *incremental* improvement, rather than absolute numbers—taken as a single snapshot. It's important that the aspects of sustainability are, for lack of a better word, *sustainable in and of themselves*.

So to set your enterprise up for success, after the assessment process, go back to Chapter 2, and reassure that your statements on sustainability, as stated there (and here again for your convenience) are as follows:

1. *Short* and *simple?*
2. *Specific* to the company?
3. *Visible* and easy to find?

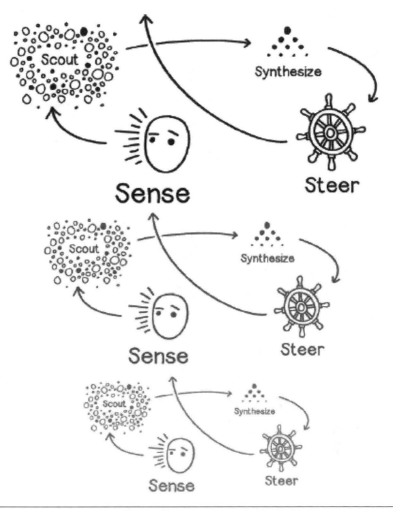

Figure 5.2   An adaptation of Plan-Do-Check-Act. (Courtesy of Dr. Cynthia Scott.)

4. *Connected* to the overall mission/vision of the particular organization?
5. *Sharing* links to relevant corporate documents for transparency (i.e., your EMP)?
6. *Tested* with employees?
7. *Tested* with suppliers and partners?
8. *Tested* with customers and other stakeholders?
9. *Evaluated* as necessary to confirm relevancy?
10. *Updated* as necessary to keep it relevant?

This is the way to keep the spiral a positive one.

Figure 5.3   The Chinese characters for *kaizen*.

Think of this as applying kaizen to your sustainability efforts. Kaizen (see Figure 5.3) is a combination of the Chinese characters for (top to bottom in the figure) "Change" and "Good," and is generally interpreted as "continuous improvement"—here applied to your enterprise's sustainability efforts.

From the Kaizen Institute,[*] we can think of this as applying these philosophies:

- Good processes bring good results
- Go see for yourself to grasp the current situation
- Speak with data, manage by facts
- Take action to contain and correct root causes of problems
- Work as a team
- Kaizen is everybody's business

This chapter has presented the transition from theory to practice—putting the tires on the ground and beginning to move. Next, we introduce the Sustainability Wheel—an instrument you can easily apply to get a read on where your enterprise is on this journey and how ready it is to accelerate.

In it, you'll see how we can measure these dimensions and provide a "signature" of your enterprise with respect to project-, program-, and portfolio-level sustainability.

---

[*] http://in.kaizen.com/about_us/definition-of-kaizen.html (accessed October 6, 2014).

6

INTERPRETING THE
SUSTAINABILITY WHEEL

Every thought you produce, anything you say, any action you do, it bears
your signature.

–Zen Buddhist Monk, Thích Nhất Hạnh

In this chapter, the rubber truly hits the road, as we take the
Sustainability Wheel™ out for a spin. We discuss pilots of the
Sustainability Wheel, initial reactions and results by pilot users, and
review and interpret many of the key *signatures* you may see—the
"Sustainability Radar™" that could arise from using the instrument.

Initial Feedback

The Sustainability Wheel has been tested with several enterprises of
varying size and focus, from a global leader in IT, to mid-size con-
sultancies, to one of the largest beverage companies on the planet,
and universally, they have told us that it has provided insight and was
"easier than expected" to complete. In verbatim comments from one
of our users, "The questions and the process of thinking about the
answers was very informative, especially the project questions. We've
been ramping up our PM training in parallel with our fledgling sustainability program and I'm seeing the clear connections now. Both
are topics of keen interest and passion."

Interpreting the Sustainability Radar™ Signatures

What we've done here is to look at the possible signatures and to
attribute a characteristic with "exaggerated" versions of the signatures
to help convey our interpretation. Since this is a visual representation
with six dimensions, there are an infinite number of permutations

along the six analog scales. We've selected a fairly large number of representative signatures to help identify the characteristics of several types, creatively named to evoke the key behaviors of the enterprise with respect to sustainability. The names associated with the signatures shown here are intentionally somewhat emotional in nature. Your signature will likely be much less extreme but may have the same *basic* shape and orientation. The idea of using these names and exaggerated signatures is not only to show—in the extreme case—what this could mean for you (either good or bad) but also to provide motivation to improve—or if you are lucky enough to be a leader—to maintain your place and improve even further from there.

In each of these signature "caricatures," we provide a paragraph or two on "what it means" and then move right into "what to do," so that you and your enterprise can consider actions and programs that could build on strengths and react to—or better yet, proactively prevent—weaknesses.

The Signatures

Leader (Strong in All Dimensions)

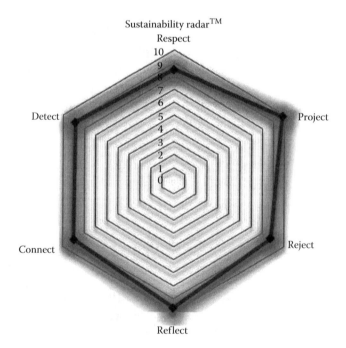

What It Means   Since this is mainly a self-scoring instrument, a signature like this could simply mean that you were "too easy of a grader." Most likely, though, it is an indicator of a well-balanced leadership position. Your enterprise's Sustainability Wheel is rolling.

What to Do   Your focus can be on maintaining your position and being exceedingly vigilant for any downward trends, by consciously reviewing these dimensions and looking for areas of improvement, even incremental improvements, in the way of "kaizen." Also, an enterprise at this level of sustainability excellence should be sharing its best practices with peers and seeking to promote their success as an example.

Laggard (Weak in All Dimensions)

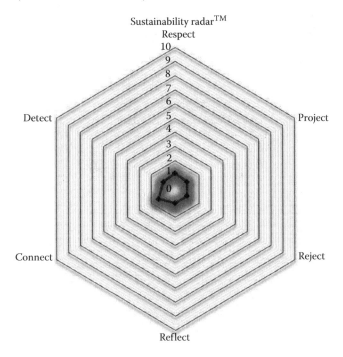

What It Means   Laggards have signatures that are centered and small—representing low scores literally all around the Sustainability Wheel. Since this is a self-scoring instrument, it could be a result of a "too-tough-on-yourself" scoring basis, or it may indeed indicate that your enterprise has a very long journey ahead to gain awareness and accomplishment of your sustainability goals.

What to Do   We suggest that you start with respect. The first step—and basis—for inflating this signature toward a leader is taking the "leadership" step of making the triple bottom line (TBL) a priority for yourself and the enterprise. This means acknowledging the value (see books like *Green to Gold*) of ecological and social accomplishments. One way to do this is to include the numbers related to this in your annual reports as some companies have done, with good results. A focus on TBL thinking will enable the other dimensions. As the hub of the wheel, a missing "respect" dimension means there is no way to make the rubber hit the road—and your Sustainability Wheel simply won't turn. Get it turning with a focus on the respect dimension, then work on reflect to get some traction. Refer back to these sections of the book for more details.

Theorist (Weak in Connect, Reflect, and Reject)

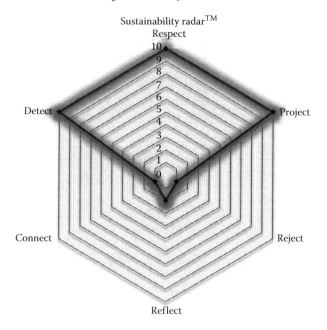

What It Means   This signature indicates high levels of respect and a strong capability of identification of threats and opportunities (in other words, all sustainability risk). However, when it comes to applying this to operations or perhaps even to the whole of the enterprise's staff, the theorist is ineffectual. It's as if the hub is turning but the wheel isn't engaged. Low scores in connect mean that the enterprise doesn't communicate with the world about their outstanding commitment to sustainability at the ideation level, employees and other stakeholders aren't aware, and on top of this, the enterprise—in general—doesn't operate efficiently, which seems counter to their lofty goals.

What to Do   The theorist needs to flex their "mission muscle." They need to reach out to—and grab—employees' and stakeholders' attention (reflect and connect), assuring that their mission is clear to all. Engaging employees with programs such as Marks & Spencer may be in order (see *Green Project Management*). And although there is always the danger of becoming a greenwasher, the theorist needs to be much less shy about getting the word out about their strong sustainability core.

Greenwasher (Weak in All Dimensions except Connect)

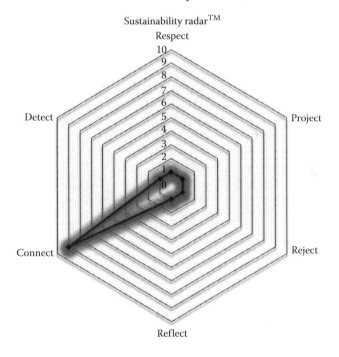

What It Means   The Greenwasher signature indicates a much stronger connect dimension than any other (particularly at the "expense" of respect and reflect). This type of enterprise is vocal (perhaps "loud" is a better way to put it) about all they do in the area of CSR, but is blind to the opportunities and threats, is poor at engaging their staff, has issues with inefficiencies and waste (internal and external), and has not taken advantage of the strength of their project, program, and portfolio management talent. They may end up with short-term economic gain from this behavior, taking advantage especially of those customers who believe their vacant CSR claims, but they will likely face the costs of poor greenality (see our coverage of that topic in our first book, *Green Project Management**) in the medium and long term.

What to Do Just as with the laggard, the first effort must be in the respect dimension, perhaps starting with an investigation of all of the claims and efforts made in what apparently has been a very effective connect campaign. The job now has to be to see what can turn these claims into reality—from perhaps somewhat random statements to a thoughtful, valid set of mission, vision, and value statements that can truly be supported. It may be very wise to start small and work your way out.

Exploiter (Weak in All Dimensions except Project)

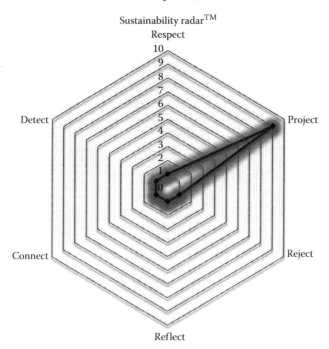

---

* Maltzman, R. and Shirley, D. Green Project Management, CRC Press, Boca Raton, FL. http://www.crcpress.com/product/isbn/9781439830017.

What It Means   Similar in several aspects to the greenwasher, the exploiter has put almost no effort into any of the dimensions except project, and although that dimension has elements of PM maturity level, we'd be able to deduce from the other behaviors indicated that part of the reason that *project* that is showing as "high" is mostly to gain advantage (opportunity) from CSR efforts, perhaps for the advertising and image "bump."

What to Do   The exploiter could start by looking at the areas in which it seeks to gain advantage and stepping back to see where the respect dimension could be enhanced. Again, respect is the dimension which is foundational to the Sustainability Wheel. Without it, signatures (and behaviors) like the exploiter can almost be expected. In this case, the enterprise can look at what it has been seeking to gain from the project dimension and work on respect as well as engaging in the other side of the project dimension where it is already strong—the project, program, and portfolio managers. It may make sense to assure that the Project Management Office (PMO) of such an enterprise is involved in planning the creation and rollout of the mission, vision, and value statements to all staff.

Drone (Weak in Respect)

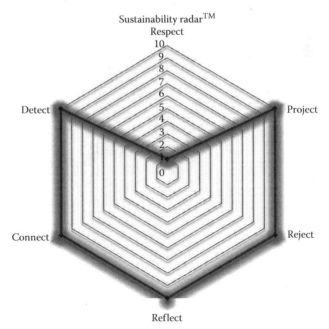

What It Means   This signature indicates leadership everywhere but in … Leadership. With this signature, or similar, there's an indication that your enterprise is clicking on all cylinders, but it's doing so without a strong sense of direction or ideation. There's no pilot. There may be a bit of luck that is resulting in the other dimensions being so well executed. The enterprise certainly cannot count on that to last. It's a bit like a drone flown by committee, and the landing will likely not be pretty.

What to Do   If your enterprise is in this mode, it may be only necessary to take what appears to be existing bottom-up commitment and adapt that as the endorsed leadership stance, with (1) an acknowledgement of what's been accomplished by the team members on their own and (2) a statement that the same level of commitment has been taken on by the organization's leadership, with senior management clearly communicating to staff and key stakeholders that they are committed to a CSR-driven enterprise.

Efficient Bamboozler (Weak in Respect, Reflect)

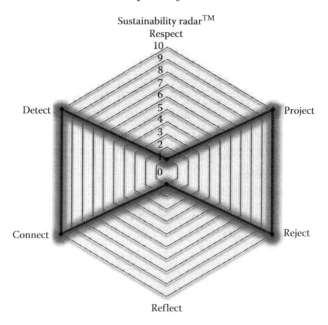

What It Means   This "bowtie" signature indicates strong risk capabilities for sustainability—both threats and opportunities as well as a potentially strong and mature PM capability. This signature also indicates a particular efficiency in operations. Further, an enterprise with this signature has done a good job convincing other stakeholders—perhaps customers—that the enterprise has a strong commitment to CSR, but that commitment truly isn't there, nor do the enterprise's employees "get it" when it comes to TBL thinking.

What to Do   The efficient bamboozler just needs to put their effort into "doing what they say they do." Already in place is a capable PM infrastructure sound risk management and a convinced customer! Now this enterprise only has to "make it real" by taking on a leadership commitment to CSR and getting that commitment conveyed to staff. The good news is that with a strong PM capability, excellent risk management, and a set of already-convinced stakeholders, the groundwork is already there for senior management.

Inefficient Optimist (Weak in Reject, Detect)

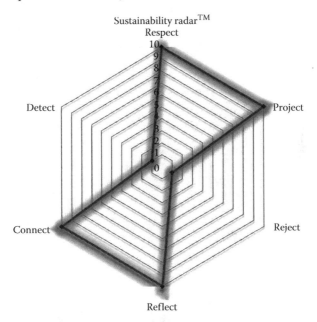

What It Means   This signature describes an enterprise that is likely relatively wasteful and inefficient in operations and that does a poor job of identifying, analyzing, and responding to CSR-related threats, although they do have a well-established CSR philosophy that has successfully reached employees and external stakeholders. There may be a correlation between the two dimensions that are lagging in this signature. Imagine that an enterprise literally is not aware of the threat of a certain type of waste (think of a leak of a mildly poisonous chemical) that it is releasing into the environment. Despite best efforts at the top levels of the enterprise and an excellent score in conveying CSR goals to staff, the poor "threat awareness" will help prevent them from even knowing that the leak is occurring.

What to Do   This enterprise's strong scores in the respect and reflect dimensions should be promoting a strong risk culture in their project management area with respect to CSR, but for some reason, that is not happening. There may be an overarching cultural issue at play, such as a strong risk-seeking culture (more common in some Asian countries). One solution or suggestion in this case is to put the strong PM capabilities to work as trainers for other managers in terms of risk management. Project managers should, through their training, have the tools to override a risk-seeking culture (at least for the sake of risk identification), and should be a receptive audience based on the strong scores in the dimensions of respect and reflect.

Shy Pessimist (Weak in Project, Connect)

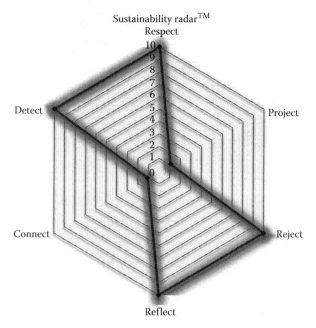

What It Means This signature indicates a very low project and connect score. We can interpret the project score in one or both of two ways—a lower maturity level for the PM community and/or an inability to identify the positives of CSR. Also, this enterprise, while stating the importance of CSR to staff, and operating efficiently, has not brought this message to other stakeholders, notably customers. There could be a correlation here, as a strong PM maturity level should yield higher stakeholder identification and engagement for the enterprise.

What to Do Similar to the inefficient optimist, it may be possible to have the project management population lead training on stakeholder management. However, in line with the enterprise hallmark of operating efficiently, it would gain a two-for-one advantage by upskilling the PM staff and general employees as well at the same time. In fact, this enterprise has the wherewithal to be a leader in sustainability if they can improve their connect and project scores and they should be able to get them both brought up simultaneously.

Shy Optimist (Weak in Detect, Connect)

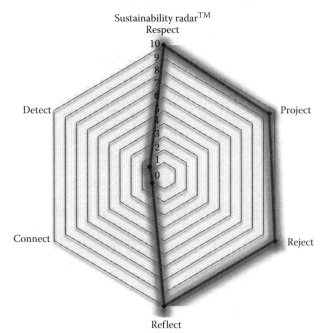

What it Means This enterprise is not very skilled at identifying CSR-related threats, although the fundamental PM capability should allow it to do so. They also have a gap in being able to communicate their CSR success to a broad set of stakeholders. There may be correlation, in that the enterprise is not considering the broad and deep set of stakeholders they may have to be blind to threats coming from those unidentified stakeholders. For example, they may have failed to identify a particular regulatory agency that is in fact charged with monitoring their behavior and outcomes. Clearly, they need to fix this oversight so that they can proactively work with this agency.

What to Do If the higher project score indicates a mature PM capability, the PMO, or PM best practices, or center of excellence (or some other central PM) organization can help build up the enterprise's capability to identify and manage stakeholders and threats. Further, if the high score in the project dimension also (or instead) represents the enterprise's ability to identify positive risk (opportunities), the enterprise needs to train those responsible for risk identification to specifically take on a "negative" mind-set when determining all that could go wrong when it comes to an enterprise's CSR implementation. This may include, for example, looking at the worst-case scenarios for something like a chemical spill, expanding the scope of the research to look for downstream populations as well as flora and fauna. It can also mean improving templates such as risk registers, to assure that they include risk categories for safety, social, and ecological risks—and that the field staff that identify risk are using these templates.

Unmoored Efficiency Expert (Weak in Detect, Respect, Project)

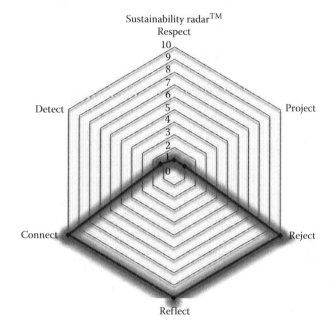

What It Means   A profile like this means that the enterprise, similar to the pilotless plane, has no overriding sense of direction from senior management with regard to sustainability policy (mission, vision, values) and beyond that, has no ability to identify opportunities or threats in this area. Depending on the specific results with project, this signature may also signify a lower maturity in project management in general.

What to Do   As in many of these cases, start with respect. A strong senior management commitment to CSR will "right" a lot of wrongs here, and likely will improve not only the capability to manage CSR risk, it will strengthen the project management community and raise their level of maturity—with a focus on the program and portfolio level—and the longer term.

Efficient Automaton (Weak in Connect, Detect, Respect)

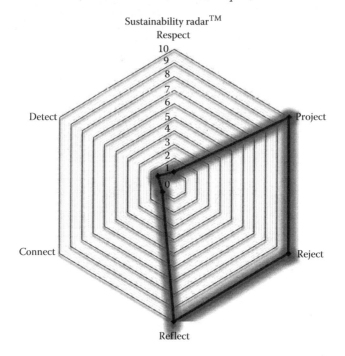

What It Means This signature is a little odd but it's certainly possible. It indicates that, like the pilotless plane, there is no fundamental direction or guidance from senior management related to CSR. The enterprise is good at identifying opportunities and may likely be mature in project management capability, and, again like the pilotless plane, has a staff that is (somehow) focusing on sustainability goals, but likely that is not going to last. The high score in reject means that the enterprise is working efficiently but again, without direction with respect to CSR.

What to Do Begin with respect, and perhaps in parallel, begin to investigate the areas of threat which may exist that have roots in CSR. After the senior managers have established a clear mission/vision statement and firm CSR values, work can begin to communicate these to other stakeholders, to gain traction with customers, and to gain other advantages of TBL thinking.

Pilotless Altruist (Weak in Respect, Project, Reject)

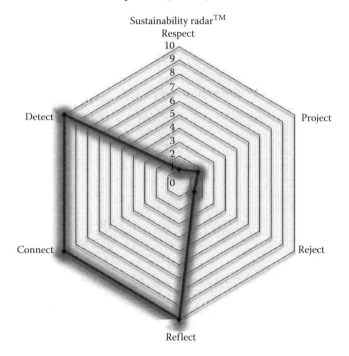

What It Means   This signature represents an enterprise that has little or no senior management direction with regard to CSR, and yet had its staff engage and look for threats related to CSR. Inefficient though they are, this enterprise has connected with external stakeholders and is trying to implement responses to CSR threats. However, they are nearly blind to the opportunities and may have a low maturity level in project, program, and portfolio management.

What to Do   Start with respect—this enterprise shows significant promise, if only the senior leadership will sign on and do a few key things (after establishing a CSR-oriented mission, vision, and values, of course). They can promote the assessment of CSR opportunities and improve efficiency–perhaps concurrent with improving PM maturity, which will help in that area. For example, senior leadership could launch a program to investigate and improve areas that have been shown to be wasteful or inefficient.

Carefully Inefficient Pilot (Weak in Project, Reject, Reflect)

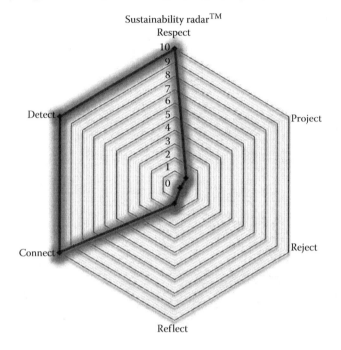

What It Means Enterprises with this signature have created CSR-oriented mission, vision, and values and are sensitive to threats related to CSR but have less-than-mature project, program, and portfolio management and are operating inefficiently. They seem to have skipped engaging their staff but have indeed done well engaging external stakeholders.

What to Do Off to a good start with solid CSR-oriented ideation, this enterprise needs to fill in several key internal gaps to make good on the promise that it has to excel. It can start by engagement of staff—including the project management population, and can expand to work on (detect and fix) operational inefficiencies and to be better at finding opportunities for gains in the area of CSR.

Inefficient Pessimist (Weak in Project and Reject)

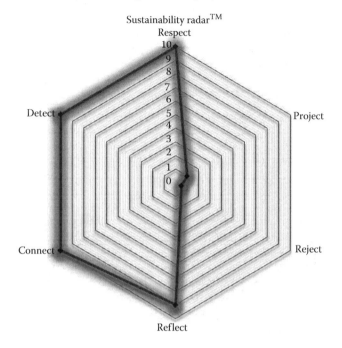

What It Means A signature such as this indicates strength in ideation as well as good traction with both staff and the outside world. However, it also indicates deficiencies in PM maturity as well as identifying areas in which sustainability could be a significant growth area. With respect to risk, this organization is focusing more on the threats—or at least is not properly identifying, analyzing, and responding to CSR-related opportunities. One other trait of this signature is that it indicates a lack of focus on efficiency. This could mean either inefficiencies in operations or in terms of general waste. Given the strong respect score, it's unlikely, but possible, that this indicates the release of toxins or waste into the environment.

What to Do The coaching here is to kick up the level of PM maturity in the enterprise. You may want to consider going for PMP® Credential levels that exceed 35% and instituting a PM methodology that is common for all projects, as well as assuring that there is a PMO in place, which is charged with providing guidance in terms of methodology and coaching and caring for the advancement of project managers in their careers. This will do two things—first, it will indeed bring up the caliber of your project, program, and portfolio management, and by doing so, you will probably simultaneously solve the weaknesses in identifying opportunity, because that's something that a mature PM culture will do well.

Planeless Pilot (Weak in Reflect)

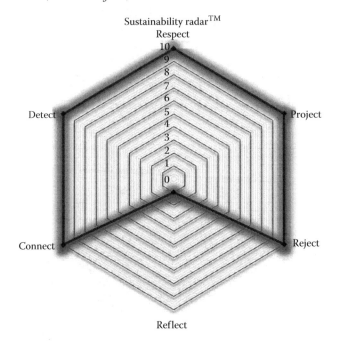

What It Means   Here, we have a signature that reflects (excuse the pun) a disability to convey the message of the importance of TBL to the staff. It's like a pilot of a large jumbo jet, flying along without a crew, passengers, or even a jet—just the pilot flying along—to the correct destination, mind you—but without the minor detail of the aircraft. Otherwise the enterprise is in pretty good shape. It's a bit unusual to have this profile because in it, the rest of the world knows about the sustainability-oriented mission, but somehow the staff didn't get brought along for the ride.

What to Do   It's pretty simple in this case—it's about change management and engagement. The enterprise staff needs to know the what, where, why, how, and when of the CSR elements of senior management's plan. They need to know that they're an important part of integrating sustainability into the fiber of the enterprise. One way to do this is to use the annual report, likely in outstanding shape to show TBL results, and assure that staff is aware of these results.

Shy Drone (Weak in Respect, Reject, Connect)

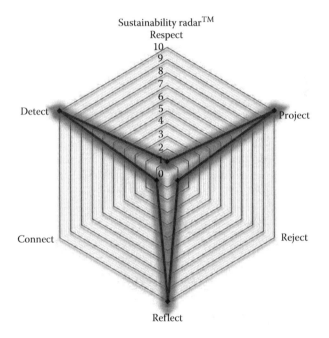

What It Means   This enterprise has strengths in dimensions that, on their own, don't have that much capability to do anything in the area of sustainability. Yes, they are capable of identifying and dealing with risks (opportunities and threats) in this area, and the staff of the enterprise has somehow been brought on board. But there is no connection to external stakeholders (that's the "shy" part), and it's very unlikely that this temporarily good situation will last.

What to Do First step: add the ideation piece we've talked about so much in this book. Mission, vision, values have to be stated. Change from a drone to a real aircraft with a pilot—one who is focused on getting the plane where it is supposed to go. Luckily, the high scores in reflect indicate that there will be a very easy time to get the "crew" on board in that they're already operating with a focus on the TBL. It will be also necessary to get a focus on efficiency and waste as that is still a weakness.

Operator (Weak in Project)

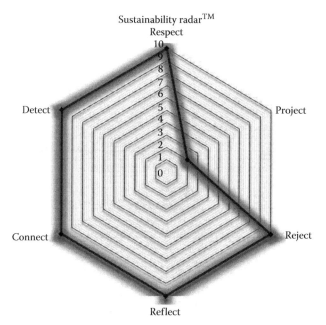

What It Means Weakness only in the project dimension could mean simply a lack of ability to identify and manage sustainability-oriented opportunity, but most likely it will reflect a deficiency in project, program, and portfolio management maturity. We use the name "Operator" to convey the idea that project management is a distinct discipline from "operations" (manufacturing, accounting, sales) and that what makes project management different and indispensable is the set of capabilities and tools that enable project managers to successfully complete projects. Those in operations do not necessarily have these capabilities. This is not to diminish operations—only to distinguish it from PM.

What to Do   As in the inefficient pessimist, we advise that you focus on increasing the level of PM maturity in the enterprise. You may want to consider going for PMP Credential levels that exceed 35% and instituting a PM methodology that is common for all projects, as well as assuring that there is a PMO in place, which is charged with providing guidance in terms of methodology and coaching and caring for the advancement of project managers in their careers. This will do two things—first, it will indeed bring up the caliber of your project, program, and portfolio management, and by doing so, you will probably simultaneously solve the weaknesses in identifying opportunity, because that's something that a mature PM culture will do well.

Theoretical PM (Weak in Connect and Reflect)

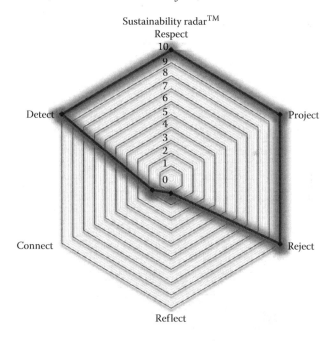

What It Means   This signature is interesting—it has strength in key dimensions, yet is lacking in critical ones that will lead to an inability to get things done. It's similar to an academic course in project management with good detail on earned value and psychological theory but nothing to teach the students as far as how projects *really* get done. The low scores in connect and reflect mean that the enterprise with this signature hasn't communicated the (well-developed) mission and vision to their staff, and they haven't made any other stakeholders aware of their will to work on the TBL.

What to Do   We suggest emulating an organization like WalMart and what they've done to improve the sustainability elements of their supply chain.* By creating the Sustainability Consortium™, they connect with their suppliers and other stakeholders, and in parallel, have communicated just how important sustainability is to the leadership with training for employees. They've created a Sustainability Hub for suppliers, where their supply chain partners can learn about how to make a difference *along with* WalMart. If you're a smaller enterprise, begin with your staff and work outward. Engage them with the mission and vision and get their buy-in—and help—in further connecting with customers, suppliers, and other stakeholders.

Fearless Leader (Weak in Detect)

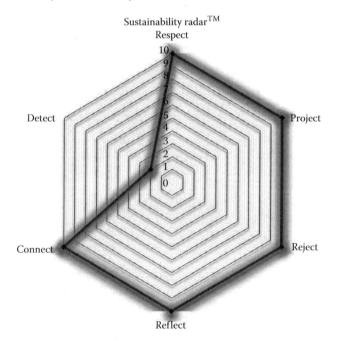

---

* WalMart corporate web page. http://corporate.walmart.com/global-responsibility/environment-sustainability/sustainability-index (accessed January 29, 2015).

What It Means   The good news here is that this signature indicates strength in all other dimensions. The weakness is in detect—the ability to identify and deal with threats to sustainability outcomes. This means the enterprise with this signature may be heading out to do all sorts of things with the best of intentions but not realizing—for example—the aspect of secondary risk, in which the risk response you've put in place causes a new threat (like an air bag that deploys and causes injuries that wouldn't have taken place otherwise). We call this signature "fearless leader" because perhaps they should have a little bit of fear—or at least caution—when moving ahead on sustainability threats and issues.

What to Do   The remedy here is twofold. First, go with your already-existing strengths in project management, which should include a solid background on risk management. With that strong framework, identifying risk should be easy and natural to you—as should the next steps of analyzing and responding properly to the threats. What may be missing is simply the connection between this classic project management capability and knowledge of the sustainability "practice area." And that brings us to the second step, which is to use benchmarking and partnering with sustainability-knowledgeable organizations. Ironically, this is a form of "risk transfer," which is one of the classic risk responses for threats.

Pessimistic Planeless Pilot (Weak in Reflect, Reject, Project)

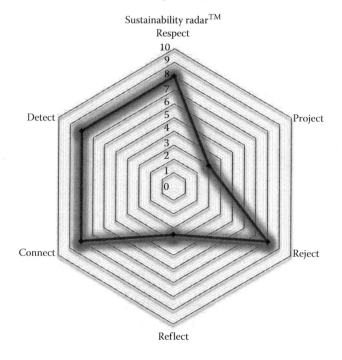

What It Means   This profile features two key weaknesses–but not at all a hopeless situation. The strengths that come with this signature are all powerful plusses:

- A solid mission
- Efficiency
- An ability to identify, analyze, and manage threats
- The means to let stakeholders know how their dedication to the TBL is an advantage

All of these will serve this enterprise well to help make up the gaps they have in getting buy-in from their staff and maturing their PM discipline (as well as finding opportunities in sustainability).

What to Do   The remedy here is twofold. Unlike the planeless pilot, the *pessimistic* planeless pilot cannot tap its existing strengths in project management, which means it must build from its strength in identifying threats to "flip to the other side" and do the same for positive risk (opportunity). The overall PM culture here may be a little weak.

So, we suggest starting with an effort to shore up the PM maturity of the enterprise. And, as in the planeless pilot, we advocate the use of benchmarking and partnering with sustainability-knowledgeable organizations. This is ironically a form of "risk transfer," which is one of the classic risk responses for threats.

Sustainability Wheel Pilot Results

Wanting to be sure that this instrument had value and ease-of-use, we tested it with several enterprises of varying discipline and size. This allowed us to make improvements in use and to look at preliminary results from these users, which we share with you in the following.

Global, Well-known IT Leader

We provided this instrument to a very large global, well-known corporation that designs, sells, and supports hardware and software services. They employ a significant project management staff. On using the Sustainability Wheel, they discovered quite a bit about themselves in the process, based on the way the questions are designed (this is intentional). Their results are shown in the following.

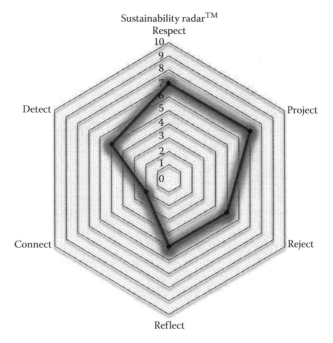

Looking at their signature, we see that the somewhat collapsed image indicates a borderline laggard signature, although the shape and fact that it's not completely collapsed indicate that it also could be a shy optimist (see description earlier). This means they may want to lean on their central PM organization (which they do indeed have) to help build up their capability to identify and manage stakeholders and threats. Further, their high score in the project dimension also (or instead) represents this enterprise's ability to identify positive risk. They could train those responsible for risk identification to specifically take on that "negative" mind-set when determining all that could go wrong when it comes to their efforts on CSR implementation and staying "true" to the TBL.

Consultancy Services Enterprise

A mid-sized environmental consultancy examined the tool and got the results shown in the following.

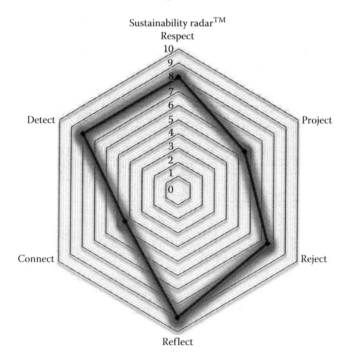

Looking at the signature of this environmental consultancy and doing a very fundamental assessment, we find that it is basically a leader (although a bit shy with the relatively weaker connect score). This is not surprising, given that it is an environmentally oriented enterprise.

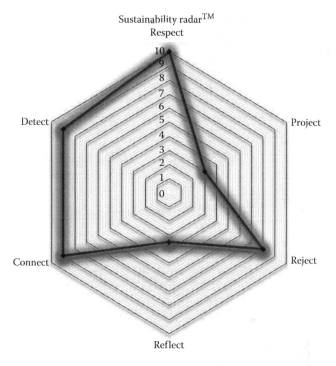

Design, Construction, Engineering Firm

A mid-sized design, construction, and environmental consulting firm used the Sustainability Wheel tool and its resulting signature is shown in the following figure.

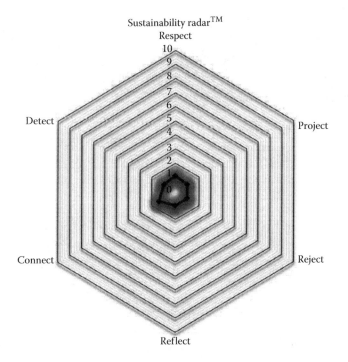

Their score indicates that they are a pessimistic planeless pilot, indicating that they could be well served by an increase in PM capability and maturity, and—importantly, need to convey their outstanding mission-level values to the full set of team members, as least as well as they have to all other stakeholders—where they excel. Further, noting that the "project" dimension has the attribute of sensing positive environmental risk, this enterprise needs to look at how social and ecological aspects of their work may be able to be viewed as and processed as opportunities, not only threats. As of this writing, this organization was actually in the process of adding several new PMP credentialed individuals to their staff and was following a path to improve the PM capability of their existing staff.

But Wait, There's More…

There are many more of these signatures than the 21 we've presented here. Also, aside from this book, we will continue to work with enterprises to help them assess their current status and areas for improvement. We encourage you to contact the authors to advise them of interesting and productive findings from the use of the instrument, especially when the findings have resulted in positive changes, reducing waste, taking advantage of sustainability opportunities, or reducing sustainability-oriented threats.

Please continue to check in at our site—earthpm.com—for the latest updates and availability of tools and results as more and more enterprises take advantage of the Sustainability Wheel.

Index

A

Accenture.com, 53
Amazon, climate counts
    scorecard, 97
American Society for the Prevention
    of Cruelty to Animals
    (ASPCA), 137–138

B

Boston's Big Dig, 14–16
BREEAM assessment, 116

C

Case study
    General Motors (*see* General
        Motors (GM))
    Marks and Spencer–beyond Plan
        A, 133–134
    Patagonia, 44–46
    Shell Oil, 132–133
    Stonyfield Farms, 46–47
    AT and T, 123–124
    Walkers Crisps, 132

Change intelligence (CQ) model
    chartered project, 30–32
    elements, styles, 27–28
    executive change leaders, 35
    executives, managers and
        supervisors, 34–35
    "high hands" styles, 28–29
    PM's adaptation of change,
        33–34
    project manager change
        leaders, 36
    project managers as *Hands*
        people, 30
    supervisor change leaders, 36
    TBL, social and ecological
        elements, 32–33
Change management, 27, 37
Circumspect
    Chinese characters, Kaizen, 146
    Plan-Do-Check-Act adaptation,
        144–145
    sustainability statements,
        144–145
    Sustainability Wheel, 146
Cisco, 66
Clean Air Act of 1970, 54

Clean Water Act of 1972, 54

ClimateCounts.org, 96

Committee of Sponsoring
Organizations of the
Treadway Commission
(COSO), 63

Compact fluorescent lamps
(CFLs), 115

Cool CO2mmute™, 70–71

Corporate social responsibility
(CSR), 26, 88, 91, 106

COSO, *see* Committee of
Sponsoring Organizations
of the Treadway
Commission (COSO)

Cost Performance Index (CPI), 9

CQ model, *see* Change intelligence
(CQ) model

Creating Shared Value (CSV)
in *HBR* article, 11
Nestle's website, 11
paradigm, 12
social and environmental
issues, 10–11

CSR, *see* Corporate social
responsibility (CSR)

CSRHub.com, 91

CSV, *see* Creating Shared Value
(CSV)

D

Dialect
communications
body language, 140
channel, 137
emotions, 137–139
innovation, differentiation,
and market share, 141
intellect, 139
language, 139
listening, 140
passion, 139

reading, 140
spectrum of reactions, 138
sustainability effort, 136, 138
TED talks, 137
PMO Symposium in 2013, 135
project and program
managers, 136

Dow Jones Sustainability™ Indices
(DJSI), 86–87

Drucker *vs.* PM view, 14

E

earthpm.com, 175

EarthPM™
assertions, *Green Project
Management*, 50–51
"hub," 51
mission, 50
objectives, 50
project management and
"green," 49
"road hazards," 51
website, 49

ecomii.com
desirability as an employer, 127
employee retention, 127
inspired problem solving, 127
stressed budget, 127

Economics, Social, Governance
(ESG) dimension, 90–91

EMP, *see* Environmental
management plan (EMP)

EMS, *see* Environmental
management system (EMS)

ENVELOLOGIC post, 62

Environmental and sustainability
education (ESE), 126–128

Environmental management plan
(EMP), 42, 44, 54–55

Environmental management system
(EMS), 55–57

EPA ENERGY STAR® Partner, 47

G

GBCA, project's success
emissions, 118
energy, 117
indoor environment quality, 117
innovation
final disposal and monitoring,
119–120
gray water collection
system, 118
indoor plants, 118
recycling and reuse, 119
waste control, 119
waste prevention, 119
land use and ecology, 118
management, 117
materials, 118
reducing organization's waste
composting, 121
electronic communication, 120
GHGs, 122
principles of sustainability, 122
products design/
manufacturing, 123
redesigning, 120–121
reducing packaging
materials, 122
transport, 117–118
water, 118
Gear Model
bottom-up processes, 8
description, 2
and OPM, 3
organizational strategy, 4–5
portfolio, 4
program, 4
"project management," 3
projects, 4
Pulse of the Profession, PMI, 5–6
SEF, 3
General Motors (GM)
employee engagement, 48

"hub," 47, 49
innovations, 48
integration, 48
mission statement, 47
planning and decision making, 48
socially responsible operations, 47
sustainability program, 49
transparency, 48
Global 100
criteria, 100
reflect dimension, 101
from website, 98, 100
weighted elements, 99
Global Initiative for Sustainability
Ratings (GISR), 89–91
GM, *see* General Motors (GM)
Good sustainability mission/vision
statement, 43–44
Greenhouse gases (GHGs), 122–123
GreenNurture program
articles, news and announcements,
RSS feeds, 84
forum, 83–84
personal home page, 83
recognition for participation, 84
video training, 84
Green Project Management, 43, 46,
49–50
*The Guide to the Project Management
Body of Knowledge
(PMBOK® Guide)*
Enterprise Environmental
Factors, 58–59
Organizational Process Assets, 57
organization's sustainability
efforts, 59
processes and procedures, 57–58

H

Harvard Business Review (*HBR*)
article, 11
Hewlett-Packard (HP), 66

Hub, sustainability wheel; *see also*
    Mission/vision statements
  description, 41
  EarthPM™, 50–51
  GM, 47, 49
  Stonyfield Farms, 47
  sustainability efforts, 41–42
  Think of the Earth, 41

I

Integrating sustainability
  bottom-up processes, 8
  NASA, 7
  "on message," 8
  project orientation, 7
  projects, strategy and steady-state
    benefits, 8–9
  project success, 9
Intellect
  benchmarking, 141–142, 144
  enterprise's sustainability, 142
  sustainability strategy, 143
  SWOT model, 143
InterfaceFLOR's senior
    management, 69
Interface Global, 66
Intergovernmental Panel on Climate
    Change (IPCC), 62, 96
International Organization for
    Standardization (ISO),
    55–57

J

Johnson & Johnson, 66

K

Kaizen Institute, 146

L

Life cycle assessment (LCA), 132

M

MAP, *see* Mission Action Plan
    (MAP), Stonyfield
Massachusetts Green High
    Performance
    Computing Center
    (MGHPCC), 65
McKinsey.com, 129
Mission Action Plan (MAP),
    Stonyfield
  challenges, 81–82
  core group of employees, 80–81
  employee engagement, 79
  environmental literacy, 80
  GreenNurture program, 83–84
  measuring results, 81
  seeking step change, 82–83
  SWOT, 80
Mission/vision statements
  disconnecting and
    reconnecting, 42
  EarthPM™
    assertions, *Green Project
      Management*, 50
    "green," 49
    hazards, 51
    mission and objectives, 50
    operational sustainability
      objectives, 51
    "Get a Good Mission
      Statement," 43
    GM's, 47–49
    good sustainability statement,
      43–44
    Patagonia, 44–46
    project management, 43
    Stonyfield Farms, 46–47
Mission Zero®, 70–71
MIT Sloan Management Review,
    133
Multilevel Project Success
    Framework, 16

N

National Aeronautics and Space
Administration (NASA),
7, 96
National Environmental Education
Foundation (NEEF),
65–66
The Natural Step™, 122, 141
Nestle's CSV paradigm, 12
Newsweek 500 key performance
indicators, 95
Nongovernmental organizations
(NGOs), 78, 91

O

Organizational project management
(OPM)
definition, 3
OPM3®, 124
project, program and portfolio, 5
Organizational project
management maturity
model (OPM3®), 124

P

Pacific sustainability index (PSI),
88–89
Patagonia's mission statement,
44–46
Perchloroethylene (PCE), 123
Personal Sustainability Project
(PSP)
challenges, 77
grassroots sustainability
movement, 76
implementation, 75–76
program design, 74–75
sustainable SVNs, 77–78
tangible results, 76–77
Plan-Do-Check-Act, 37, 145

PMhut.com, 102
PMI, *see* Project Management
Institute (PMI)
PMI's OPM3®, 3
PM maturity, *see* Project
management (PM)
maturity
PMO, *see* Project/Program
Management Office
(PMO)
PMP® Credential levels, 163,
167, 174
Portfolio
alignment, 8
bottom-up processes, 8
definition, 4
Gear Model, 4–5
level, 59–60
managers, 111
project success, 12
PPM, *see* Project portfolio
management (PPM)
Program; *see also* Portfolio
alignment, 8
continuum, 21
definition, 4
Gear Model, 2, 5
and incentives, 63
level, 60–61
Plan A, 134
PMO (*see* Project/Program
Management Office
(PMO))
project success, 12
PSP, 74
sustainability, 49
Programme for the Endorsement
of Forest Certification
(PEFC™), 121
Project Management Institute
(PMI), 124–125
practice guide, 4
projects, 4

project success, 12
Pulse of the Profession study
    in 2013, 21
    in 2014, 5–6
Project management (PM) maturity
    encourage a culture of change, 23
    Multilevel Project Success
        Framework, 16–17
    and project success (*see* Projects)
    SUKAD model, 17–19
Project portfolio management
    (PPM), 60
Project/Program Management
    Office (PMO)
    "3-Click Challenge," 24–26
    continuum, 21–22
    in driving business outcomes, 23
    focus on critical initiatives, 22
    foster talent and capabilities,
        22–23
    Gantt Chart and WBS, 1, 135
    instituting smart and simple
        processes, 22
    necessary evolution, 23
    on strategy implementation, 21
Projects
    alignment, 8, 21
    bottom-up processes, 8
    definition, 4
    Gear Model, 2, 5
    level
        connection to sustainability
            mission/vision, 61
        COSO, 63
        ENVELOLOGIC post, 62
        project management, 61
        sustainability risk
            consequences, 61–62
    program and portfolio, 4
Projects success
    equation, 13
    generating value, 13
    long-term benefit realization, 8–9

"perceived success," 12
and PM maturity (*see* Project
    management (PM)
    maturity)
project *vs.* project management
    success, 14–15
SUKAD's model, 18–19
with triple bottom line, 16
Project *vs.* project management
    success, 15
PSP, *see* Personal Sustainability
    Project (PSP)
Purpose, 38–39, 107

R

Road
    circumspect
        Chinese characters, Kaizen,
            146
        Plan-Do-Check-Act
            adaptation, 144–145
        sustainability statements,
            144–145
        Sustainability Wheel, 146
    dialect (*see* Dialect)
    dimensions, 135
    intellect
        benchmarking, 141–142, 144
        enterprise's sustainability, 142
        sustainability strategy, 143
        SWOT model, 143
RobecoSAM, 86
Roberts Environmental Center
    (REC), 88

S

Science Technology Engineering and
    Mathematics (STEM), 124
SEF, *see* Stanford's Strategic
    Execution Framework
    (SEF)

"Sense, scout, synthesize and steer" model, 37–38
SIA, *see* Subaru of Indiana Automotive (SIA)
"SMART" goal framework, 75
Socially responsible investment (SRI), 91
Socially responsible operations, GM's, 47
Spokes
  Accenture.com and their report, 53
  challenges, 71
  education in employees, 72
  EMS and ISO 14001, 55–57
  enterprise level, 57–59
  environmental management plan, 54–55
  face-to-face meetings with employees, 69
  Gallup Organization, 70
  InterfaceFLOR, 69
  MAP (*see* Mission Action Plan (MAP), Stonyfield)
  measuring results, 71
  "Mission Zero®," 70
  organizational artifacts, 53
  portfolio level, 59–60
  program level, 60–61
  project level
    connection to sustainability mission/vision, 61
    COSO, 63
    ENVELOLOGIC post, 62
    project management, 61
    sustainability risk consequences, 61–62
  PSP (*see* Personal Sustainability Project (PSP))
  sustainability 360 lives, 73
  sustainability programs and incentives (*see* Sustainability programs)

synopsis, 72–73
tangible results, 70–71
by Wal-Mart associates, 73
Stanford's Strategic Execution Framework (SEF), 3, 38
Stonyfielders Walking Our Talk (SWOT), 80–81
Stonyfield's mission statement, 46–47
StrengthsFinder survey, 70
Strengths/weaknesses/opportunity/threats (SWOT) model, 143
Subaru of Indiana Automotive (SIA)
  American Forest and Paper Association, 121–122
  CFLs, 115
  electronic communication, 121
  emissions, 118
  energy, 117
  GHGs, 122–123
  indoor environment quality, 117
  innovation
    final disposal and monitoring, 119–120
    gray water collection system, 118
    indoor plants, 118
    recycling and reuse, 119
    waste control, 119
    waste prevention, 119
  land use and ecology, 118
  lubrication system, 115
  management, 117
  manufacturing process, 115
  materials, 118
  The Natural Step™, 122
  organization, 115–116
  recycling, 114, 121
  transport, 117–118
  water, 118
Substitutability IQ format, 94

Success-Unique-Knowledge-
Attitude-Development
(SUKAD) model, 17–19
Sustainability breakdown
structure, 9–10
Sustainability IQ matrix, 93
Sustainability Perception Score
(SPS), 92, 94
Sustainability programs
above and beyond, 65
Cisco, 66
collaboration/engagement, 65
elements, 64
energy/greenhouse gas emissions
reduction, 65
Green Carpet Award, 64
Harvard project, 65
Hewlett-Packard (HP), 66
innovative/creativity, 64
Interface Global, 66–67
Johnson and Johnson, 66
MGHPCC, 65
Mid-Course Correction, 68–69
NEEF, 65–66
NEFF, 67
practice, 63
Ray Anderson's vision, 68
replicable models, 64–65
Stonyfield, 66–67
synopsis, 68
Wal-Mart, 67
waste reduction, 65
water reduction, 65
Sustainability Radar™ signatures
carefully inefficient pilot, 162
drone, 153–154
efficient automaton, 160
efficient bamboozler, 154–155
"exaggerated" versions, 147–148
exploiter, 152–153
fearless leader, 168–169
greenwasher, 151–152

inefficient optimist, 155–156
inefficient pessimist, 163–164
laggard, 149–150
leader, 148–149
operator, 166–167
pessimistic planeless pilot,
170–171
pilotless altruist, 161
planeless pilot, 164–165
shy drone, 165–166
shy optimist, 157–158
shy pessimist, 156–157
theoretical PM, 167–168
theorist, 150–151
unmoored efficiency expert, 159
Sustainability Reality Score (SRS),
91–92
Sustainability stakeholders
advocates, 105
champions, 105–106
Change Quotient (CQ), 111
confident fans, 104
corporate social responsibility,
106
Deepwater Horizon platform,
109, 111
essential elements, 107
Federal U.S. investigation, 111
followers, 104–105
"Hands" people, 113
"Head" people, 113
"Heart" people, 113
idle stakeholders, 103–104
intra-organizational conflict, 108
oil from BP's Macondo Well, 110
oil slick, 108–109
parasites, 103
poor customer service, 108
poor quality, 108
power level *vs.* sustainability
support level, 102–103
predators, 103

Project Stakeholder
Management, 102
risk register, Macondo Well,
108, 112
shy fans, 104
supporters, 105
surface slick, 109
Sustainability Wheel Pilot
consultancy services enterprise,
172–173
design, construction and
engineering firm, 173–174
earthpm.com, 175
global, well-known IT leader,
171–172
Sustainable Value Network (SVN),
67, 74, 77–78
SWOT, *see* Stonyfielders Walking
Our Talk (SWOT);
Strengths/weaknesses/
opportunity/threats
(SWOT) model

T

TBL, *see* Triple bottom line
(TBL)
Technology, entertainment and
design (TED), 137
Think of the Earth, 41
Tire
business cases, 126
Claremont-McKenna's Roberts
Environmental Center
Pacific Sustainability
Index, 88–89
ClimateCounts.org, 96
CSRHub.com, 91
detect dimension, 101
DJSI, 86–87
ESE, 126–128
GISR, 89–91

Global 100
criteria, 100
reflect dimension, 101
from website, 98, 100
weighted elements, 99
Newsweek Green Rankings,
93–96
partnering, 128–129
profit, 128
project
maturity models, 124–125
"sustainable" organization and
opportunities, 124
reject, 113–114
resources, 96–98
SIA (*see* Subaru of Indiana
Automotive (SIA))
significant breadth, Dow Jones
sustainability index, 87
strategic value creation
benefits and costs, 129–130
collaboration and sharing, 130
greenwashing, 131
issue identification, 130
propensity, 131
reputation capital, 130–131
stakeholder trust, 131
sustainability efforts, 85
sustainability leadership report
challengers, 92
IQ matrix, 92–93
laggards, 92
leaders, 92
promoters, 92
SPS, 92
SRS, 91
sustainability stakeholders'
identification and
analysis (*see* Sustainability
stakeholders)
Triple bottom line (TBL), 164,
166, 170

ecological and social
    aspects, 20, 32
economic, ecological and social, 9
"leadership" step, 150
PMOs, 24
project success, 15–16

U

Unilever, climate counts
    scorecard, 98

The U.S. Leadership in Energy and
    Environmental Design
    (LEED) certification, 116

W

Wal-Mart, 67
Water efficiency labeling standards
    (WELS), 118
Work Breakdown Structure
    (WBS), 1, 9, 135